SpringerBriefs in Geography

More information about this series at http://www.springer.com/series/10050

William Holden • Kathleen Nadeau
Emma Porio

Ecological Liberation Theology

Faith-Based Approaches to Poverty and Climate Change in the Philippines

 Springer

William Holden
Department of Geography
University of Calgary
Calgary, AB, Canada

Kathleen Nadeau
Anthropology Department
California State University
San Bernardino, CA, USA

Emma Porio
Department of Sociology and Anthropology
Ateneo de Manila University
Quezon City, Philippines

ISSN 2211-4165   ISSN 2211-4173   (electronic)
SpringerBriefs in Geography
ISBN 978-3-319-50780-4   ISBN 978-3-319-50782-8   (eBook)
DOI 10.1007/978-3-319-50782-8

Library of Congress Control Number: 2016961308

Printed on acid-free paper

This Springer imprint is published by Springer Nature
The registered company is Springer International Publishing AG
The registered company address is: Gewerbestrasse 11, 6330 Cham, Switzerland

This book is dedicated to all those actively
non-violent and peaceful foot soldiers
on the ground working with the poorest
of the poor for a more equitable
and green Philippines.

Contents

Chapter 1
Introduction

This book focuses on issues of climate change related efforts to rebuild communities of resilience in areas affected by human induced disasters and natural hazards, and increasing vulnerabilities, resulting from mal-governance and unsustainable development in the Philippines. It considers some of the responses of some of the front-line church-based social action networks and organizations working in the face of climate change and some of its disastrous impacts on the Philippines, from the perspective of ecological liberation theology. Ecological liberation theology considers the relationship between poverty, ecological devastation, and oppression as an interrelated structural problem. By taking a bottom-up approach to redressing structural quandaries, church workers involved in this movement organize faith-based communities to transform local political, social, and ecological relationships by empowering the poor. The argument is that when the affected community trusts the community facilitator, who actively works with them to devise solutions and reach their goals, from the beginning to the completion of the project, they work more effectively together as a collaborative community for their collective interests. This type of development model is different from that of the acquisitive development and globalization model that is the dominant approach of many local and international aid organizations today.

Since the late twentieth century to present, successive Philippine governments have embraced neoliberalism, which Harvey (2006, p. 145) defines as "a theory of political economic practices which proposes that human wellness can best be advanced by the maximization of entrepreneurial freedoms within an institutional framework characterized by private property rights, individual liberty, free markets, and free trade." This top-down approach has created a local environment of unregulated, unplanned, development, without implementing safeguards for the protection of poor rural and urban communities and the natural environment. The Philippines, currently, is a country with great educational disparities and income inequalities. The few are super rich, while many live in conditions of abject poverty. According to UNESCO (2015), one out of every four primary school children, today, drops out of school, before they reach the fifth grade. The CIA Fact Book (2015) reports that

© The Author(s) 2017
W. Holden et al., *Ecological Liberation Theology*, SpringerBriefs in Geography,
DOI 10.1007/978-3-319-50782-8_1

26% of the local population lives below the poverty line. Poor communities are most at risk for succumbing to climate change-related disasters. They are more prone to floods from rising sea levels, storms surges, and coastal inundations and other natural-and human-made disasters. Porio (2012, p. 3) explains, high-risk sites such as informal settlements along the coasts and riverine systems also suffer from very poor quality housing and absence of social services and badly needed infrastructure support. Also, local disaster management officials are charged with assessing damage and losses from typhoon and flood impacts: but, hardly any assessment is done at the community level, and more importantly, from the perspective of marginal and faith-based communities.

In Chapter 2, we provide an overview for understanding the current economic and ecological crisis in the Philippines. Chapter 1 looks at the relationship between poverty and oppression and how this has directly impacted the natural environment. Our basic argument is that pre-colonial Philippine communities were better positioned to resist climate-change disasters, due to having a natural environment covered with trees and all the diverse ecology that nature brings to bear, which keeps soils intact and, wherein, local homes and communities are made using natural materials attuned with the local environment. Why is this important? To rectify some of the most serious problems posed by climate change, today, local people need to have the impetus to begin thinking about creating sustainable communities that can withstand some the horrific effects of increasing typhoons. Our argument is that some viable and culture based solutions to contemporary problems of climate change can be found by looking back and implementing some of the approaches used in the past that worked.

Chapters 3 and 4 present the critical definition and discussion of climate change offered by scientific studies and real case scenarios on how climate change has impacted and continues to adversely affect the Philippines in the twenty-first century.

In Chapters 5 and 6 we discuss and analyze the history of the development of neoliberal development policies in the Philippines and why neoliberalism is currently failing to address some of the most serious problems of climate change. This chapter, then, raises questions such as what are the different types of development approaches being used? It compares and analyzes select case studies based on our fieldwork to make a case for the bottom-up ecological liberation theology model as one of the more effective development models based on our findings.

Chapters 7–9, elaborate in detail the history of the rise of the Philippine liberation theology movement and its real world applications on the ground. We give specific examples of community organizers partnering with local communities affected by climate change disasters and how they successfully met some of the difficult challenges in the process of finding ways to rebuild their communities.

Finally, Chapter 10 concludes that it is impossible to design and implement effective disaster prevention plans, without taking into consideration the perspectives of the most adversely affected communities such strategies are intended to serve. An alternative course of action would be to include locally committed and engaged intellectuals on disaster assessment teams. They are well-positioned partners with local communities to design feasible and environmentally friendly plans. Our book focuses on these reconstruction efforts by comparing them to top-down development projects in the Philippines. The research reported in this brief is based largely on our respective participant observations and fieldwork in the Philippines.

References

Central Intelligence Agency. (2015). Philippines. *The world factbook*. Retrieved from https://www.cia.gov/library/publications/the-world-factbook/geos/br.html

Harvey, D. (2006). Neo-liberalism as creative destruction. *Geografiska Annaler: Series B: Human Geography, 88*(2), 145–158.

Porio, E. (2012). *Climate change adaptation in metro manila: Community risk assessment and power in community interventions*. Paper presentation at the International Sociological Association (ISA), Buenos Aires Forum, Research Committee on Clinical Sociology (RC 46) Sessions on Essentials of Community Intervention (August 1–5).

UNESCO Institute of Statistics. (2015). *Fixing the broken promise of education from the global initiative on out-of-school children*. Retrieved from uls.unesco.org

Chapter 2
The Philippines: Understanding the Economic and Ecological Crisis

The Philippines experiences frequent earthquakes and typhoons, due to its location on the western rim of the Pacific Ring of Fire, a U-shaped series of more than 450 volcanoes in the Pacific Ocean that trace an arc along the coasts of South America, North America, Asia, and Australia. Just off the coast of Vietnam and China, south of Taiwan, and north of Borneo and Indonesia, in the South China Sea, it consists of, approximately, 7100 islands. The Philippines has a total landmass of 115,124 mile^2, 298,170 km^2, and current estimated population of 100,096,496 people (Worldometers 2015).

In the twenty-first century, the Philippines ranks number two in the world for being a country most at risk to climate-related disasters (World Risk Index 2014). Over the years, the islands have become more vulnerable and less prepared to deal with and prevent typhoon related disasters. Climate change is precipitating a new wave of super typhoons that are increasing in magnitude and occurring more frequently and unexpectedly from different pathways than before. Also, however, in large part, climate change related disasters are caused by mal-development practices and the misuse of local resources that continues to eradicate the once ecologically and biologically rich protective forest cover that, in the past, better buffered local communities from strong winds, waves, and rains. The natural and organic fertility of the soils in the earlier, by-gone, ecosystem had more deeply rooted plants and trees that played an immensely important role in preventing disasters from typhoons. Deep roots of plants and trees help to keep the soils in place, which effectively prevents flash floods and landslides, during the typhoon season. Before colonialism and modern day globalization-and neo-liberalization, the Philippines, as a whole, was better prepared to successfully adapt to weather changes.

Historically, before Spanish colonialism, the Philippines were interlinked into a pan-Asian trade network. It is well documented that Arabs, Indians, Javanese, Siamese, Sumatrans, and Chinese traded with the islanders in pre-colonial times (Nadeau 1995, pp. 44–45 and references cited). Colonization by Spain (1565–1898), the United States (1898–1946), Japan (1942–1945) and, later, the dominant Filipino elite, resulted in the total demise of the earlier tributary economy, incipient rise of

© The Author(s) 2017
W. Holden et al., *Ecological Liberation Theology*, SpringerBriefs in Geography,
DOI 10.1007/978-3-319-50782-8_2

mercantile capitalism, and continued the penetration of the modern day neoliberal capitalist economy in the Philippines. The earlier economy existed in opposition to capitalism. It was based on use-value as opposed to exchange value. Surplus was produced, but only in the sense of an excess of goods normally used for consumption being set aside for appropriation and circulation; surplus circulated on the basis of its use value (tribute) rather than exchange value (sale for profit). The pre-Hispanic Philippines hosted a rich and lush tropical forest teeming with wild life and-aqua marine and riverine life forms. IBON Databank and Research Center (2006, p. 4) reports, at least, 90% of the pre-Hispanic Philippines was covered with forest. The sudden reduction of trees resulted from the Spanish colonizers cutting logs to build ships for the Galleon Trade, churches, and forts. The tearing down of forests also made way for the plantation system-and large-scale agricultural production of crops such as fast growing trees, sugar cane, rubber trees, oil palms, pineapples, bananas, and other fruits for commercial purposes. By the time Spain ceded the Philippines to the United States only some 70% of the country still had a forest cover (Bautista 1990).

However, Reynaldo Raluto (2015, p. 25), who is partially credited for this discussion of the denuding of the Philippine forests, explains that the sudden and most dramatic reduction of woods came about when the United States colonized the Philippines. The American colonial government designated some 84% of the country's forests as part of the public domain, without regard for the local people, whose identities and livelihoods were intertwined with the life flow and creatures of the wilderness and its rivers and seas. The American colonial forest policy was to rapidly and mindlessly cut as many trees as possible for export and sale on the international market. The colonial government, explains Raluto, was mainly interested in cutting hardwoods, especially mahoganies, to meet the high demand for hardwood that was fueling the construction industry in the United States. Within a short span of 20 years, between 1900 and 1920, the remaining Philippine forests covered only 60% of the total landmass. The American colonial government then began to issue Timber License Agreements to private corporations and individuals, who were given exclusive logging rights to large forested areas for 25 years, renewable for another 25 years. Rather than replanting tree seedlings afterwards, the colonial government opened the cleared spaces up for homesteaders to farm. The American colonial administration offered to give away to disgruntled and potentially rebellious landless laborers, who together with their families, had been displaced and disenfranchised by the introduction of capitalist agricultural production on the northern island of Luzon, and elsewhere, a piece of land to homestead for 'free' on the southern island of Mindanao, which was undergoing rapid deforestation and so-called development. However, these land grants for homesteaders came with the cost of pushing the original local indigenous people and Islamic communities off their ancestral and ancient lands. This initial taking away of the land from the local indigenous and Islamic communities in Mindanao, by giving it away to in-coming Christian homesteaders, has been a leading source of a long and tumultuous history of conflicts between Muslims and Christians, which is not based on religion, into the twenty-first century. By 1950, the Philippine forests were reduced to 50% of the total landmass. Raluto (2015, p. 26) documents that this drop falls below the minimum threshold of 54% forest cover needed by a mountainous country to maintain an ecologically habitable environment.

During the postwar years (1950–1973), big logging concessions became more technologically advanced and systematically cut the remaining forests. According to the *United Nation's Food and Agriculture Organization*, in 1963, only 40% of the Philippines' total land area was covered in forest. Then, the dictator, President Ferdinand Marcos (1965–1986), gave out Timber License Agreements to reward his cronies. Although he temporarily banned logging in 1975, in response to the increase in landslides and natural disasters that were occurring when heavy winds and rains swept across the country, this was short-lived. His cronies complained, so Marcos gave in by reinstating a Selective Timber License Agreement that gave him the power to selectively allow those in favor, to continue large-scale logging and mining operations. Many of the human and environmental tragedies that happened, during the Marcos era, then as now, could have been avoided if development practices were regulated. Tragedies caused whenever mudslides flow quickly down bare hills, over the surrounding communities, could be easily prevented, if investment agents and developers were not allowed to rambunctiously bulldoze the mountains and hills of vegetative cover, while dislodging human communities from their land and livelihoods. The rapid depletion of forested areas continues to seriously concern today's environmentalists and conservationists. The Philippine wilderness areas with ecologically diverse and wild animal, insect, and plant species are in danger of becoming totally extinct. They continue to be destructively diminished by on-going legal and illegal logging and mining concessions that pollute the environment and cause violent conflicts over land rights. The forests also are disappearing as a result of the increase in tree plantations, especially palm oils, and overpopulation (Butler 2014). This on-going clearing away of everything green in the Philippines is aggravated and encouraged by current unsustainable construction and mal-development trends. Remaining green zones around cities and towns are being cleared faster, of all vegetative life, as new housing suburbs and fancy gated communities, with large areas of rolling golf courses and recreation resorts for the better off, are becoming increasingly popular in a contemporary Philippines that is, at the same time, stricken by poverty and oppression, water, land, and air pollution. The national capital of Metro Manila, for the most part, is nearly treeless, and what little green areas remain are covered in soot, as mega shopping malls, old historic buildings, new skyscrapers, and makeshift housing structures keep rising upwards with air conditioner boxes, protruding from windows, over bare, congested, and badly polluted thoroughfares. The present Philippines no longer exports hardwood on the international market. During the last decade of the previous twentieth century, Nadav (1994) reported that were no significant mahogany and hardwood forests left. The *Philippine Forest Management Bureau* and the *World Food and Agricultural Organization* estimates that the once rich forests now cover less than 24% of the country's total land area. As Raluto (2015, p. 4, 27) rightly argues, the on-going destruction of the last remaining forests and poverty are the real culprits that place the Philippines in danger of climate change disasters.

Currently, many local citizens live in extreme conditions of abject poverty. In 2006, the archipelago's official poverty rate stood at 33% but in rural areas the poverty rate stood at 46% and 71% of all poor people living in rural areas (World Bank 2010). The rural population is heavily dependent upon subsistence aquaculture and subsistence agriculture and 50% of those engaged in the former live in poverty while 44% of those engaged in the latter live in poverty (DENR Climate Change Office 2010). These people have livelihoods requiring access to natural resources, such as fertile soil and clean water, and environmental deterioration will diminish their ability to meet their basic needs; as Broad and Cavanagh (1994, p. 814) wrote:

> To live, poor people eat and sell the fish they catch or the crops they grow- and typically those who manage to subsist in this way do so with very little margin. Natural resource degradation often becomes a threatening crisis- a question of survival

There is a strong link between poverty and environmental degradation, which makes poor people more vulnerable to climate-change related disasters. Consider, for example, the Municipality of Governor Generoso, in Davao Oriental, which has 45,000 people living on only 7000 km^2 of land suitable for agriculture (De La Cerna 2005; interview). In Governor Generoso, 58% of the people live in poverty and the population overwhelmingly consists of subsistence farmers and subsistence fishers. As Jerry De La Cerna, Mayor of Governor Generoso from 2004 to 2007, stated, "We get our bread from the ocean and we get our bread from the land; we should protect the ocean and we should protect the land" (De La Cerna 2005; interview). Any form of environmental degradation that disrupts access to natural resources will thrust the poor from subsistence into destitution. Poverty and environmental degradation combined puts the poor at greater risk than better off people, by making them more vulnerable to experiencing the direct impact of disasters when climate change related calamities happen to occur.

References

Bautista, G. (1990, March). The forestry crisis in the Philippines: Nature, causes, and issues. *The Developing Economies, 28*, 67–94.

Broad, R., & Cavanagh, J. (1994). *Plundering paradise: The struggle for the environment in the Philippines*. Berkeley: University of California Press.

Butler, R. (2014). Philippines. Rainforest country profiles. Mongabay.com.

Department of Environmental and Natural Resources Climate Change Office (DENR). (2010). *The Philippine strategy on climate change adaptation*. Quezon City: Department of Environmental and Natural Resources Climate Change Office.

De La Cerna, J. (2005). Mayor, Municipality of Governor Generoso, Province of Davao Oriental, Personal Interview, Municipality of Governor Generoso, Philippines, 25 May 2005.

IBON. (2006). *The state of the Philippine environment 2006* (3rd ed.). Quezon City: IBON Books.

Nadeau, K. (1995). *Ecclesial community movement in Cebu: The Philippines*. A dissertation presented to Arizona State University in partial fulfillment of the requirements for the Degree, Doctor of Philosophy.

Nadav, M. (1994, September/October). Tropical timber: Seeing the forests for the trees. *Environmental Building News, 3*(5).

Raluto, R. D. (2015). *Poverty and ecology at the crossroads, towards an ecological theology of liberation in the Philippine context*. Quezon City: Ateneo de Manila University Press.

United Nation's Food and Agriculture Organization. (1963). Retrieved from www.fao.org/docrep/016/ap651e/ap651e.pdf

World Bank. (2010). *Poverty and equity, Philippines*. Retrieved from Povertydata.worldbank.org

Worldometers. (2015). *Philippines population*. Retrieved September 22, 2016, Retrieved from http://www.worldometers.info/world-population/philippines-population

World Risk Index 2014 in Alliance Development Works. (2014). *World Risk Report 2014*. Berlin: Alliance Development Works.

Chapter 3
Climate Change: A Conceptual Framework

Climate change, or global warming, is caused by increasing levels of greenhouse gases, such as carbon dioxide (CO_2), methane, or water vapor, in the Earth's atmosphere. As solar radiation enters the Earth's atmosphere these gases cause the sun's heat to remain within the atmosphere instead of radiating back out into space and, as a result, the Earth warms. Although CO_2 is the weakest of the various greenhouse gases it is also the most common and, consequently, is the greenhouse gas most responsible for global warming as there has been a 42% increase in atmospheric CO_2 since 1800 and a 24% increase since 1959 (De Buys 2011). A basic description of the physics of climate change is that provided by De Buys (2011, p. 10):

> Greenhouse gases trap more of the heat that Earth would otherwise radiate back into space. The retained heat charges the atmosphere and oceans- the main drivers of the planetary climate system- with more energy, loading them with more oomph to do the things they already do, but more powerfully than before.

There is no doubt remaining as to the validity of scientific studies indicating that the world's climate is warming; as Pope Francis wrote, "a number of scientific studies indicate that most global warming in recent decades is due to the great concentration of greenhouse gases released mainly as a result of human activity" (Laudato Si 2015, p. 19). "The world's foremost authority on climate change is the Intergovernmental Panel on Climate Change (IPCC), an institution," which De Buys (2011, p. 27) described as being "charged with issuing periodic assessments of the status of climate change research and the predicted effects of climate change on global society and the environment." According to the IPCC (2013, p. 2), "Warming of the climate system is unequivocal, and since the 1950s, many of the observed changes are unprecedented over decades to millennia." "Although anomalies and uncertainties will always exist," stated De Buys (2011, p. 61), "the case for a warming climate is about as solid as any scientific case will ever be." If there have been warmer climates in the past, as the paleoclimatic evidence indicates, these will pale in comparison to what is expected in the future. According to Crowley (2000, p. 276), "the warming over the past century is unprecedented in the past 1000 years."

© The Author(s) 2017
W. Holden et al., *Ecological Liberation Theology*, SpringerBriefs in Geography,
DOI 10.1007/978-3-319-50782-8_3

"Climate change," wrote Pope Francis, "represents one of the principal challenges facing humanity in our day" (Laudato Si 2015, p. 20). Addressing the causes and impacts of climate change has, in the words of Haynes and Tanner (2015, p. 358), "emerged as a major global challenge of the twenty-first century." The scope for anthropogenic climate change is so serious that even if all CO_2 emissions were to completely stop in the year 2100, global warming would still continue until the year 3100 (Gillett et al. 2011). Today, CO_2 levels in the Earth's atmosphere already exceed 440 parts per million and they continue to rise by about two parts per million every year (Ellwood 2014). If no efforts are made to reduce greenhouse gas emissions and if atmospheric CO_2 emissions are allowed to increase up to 1000 parts per million by the year 2100 the results will be, in the words of Schneider (2009, p. 1104) "catastrophic." According to Schneider (2009, p. 1104), with atmospheric levels of at CO_2 at 1000 parts per million "many unique or rare systems would probably be lost, including Arctic sea ice, mountain-top glaciers, most threatened and endangered species, coral-reef communities, and many high-latitude and high-altitude indigenous human cultures." One of the most problematic aspects of climate change are the extreme weather events that accompany it. "There is growing evidence," stated Walch (2014, p. 40), "that climate change is increasing the intensity and frequency of natural disasters, particularly hydrologic and climatological ones such as floods, cyclones, and droughts."

Increased Drought with Climate Change

One of the most profound implications of climate change is increased drought. In arid areas already susceptible to drought this natural hazard will only get worse because there will be more evapotranspiration. This means there will be more *evaporation* (the conversion of liquid water into water vapor) and there will also be more *transpiration* (the release of water from plant leaves). With higher temperatures there will be more evapotranspiration and, unless precipitation rises more quickly than evapotranspiration (a doubtful proposition in an arid area), potential evapotranspiration will outstrip precipitation thus creating drought. According to Overpeck and Udall (2010, p. 1642) "substantially more severe warming and drying lies ahead" and "global warming theory and modeling has projected an increase in drought frequency." Trenberth et al. (2004, p. 27) stated, "It is expected that unanticipated drought will be a feature of climate in the near future, particularly given continued global warming." "A basic prediction of climate science," wrote Romm (2011, p. 450), "is that many parts of the world will experience longer and deeper droughts, thanks to the synergistic effects of drying, warming, and the melting of snow and ice."

Increased Heavy Rainfall Events

Even though climate change may also lead to more droughts it may also, at the same time, generate an increase in heavy rainfall events because warmer air holds more water vapor (Combest-Friedman et al. 2012; Knutson et al. 2010; Min et al. 2011; Thomas et al. 2013; Villarini et al. 2013; Westra et al. 2013; Yang et al. 2013). According to De Buys (2011, p. 28) this will lead not only to wet places getting wetter but to "wet *events* getting wetter: storms will carry more moisture, and when they let it go, the impacts will be greater."

Sea Level Rise

One of the most certain consequences of global warming is a rise in the sea level of the world's oceans and this will occur for two reasons: first, the thermal expansion of water (as water warms it occupies more space); second the melting of the Antarctic ice sheet, and the Greenland ice sheet (Bellard et al. 2014; Overpeck et al. 2006). The melting of Arctic sea ice will not cause sea levels to rise because it produces just enough water to occupy the space it was already occupying in the sea (Chivers 2011). Overpeck et al. (2006) estimate that by 2100 sea levels could be between 4 and 6 m above current levels. "A rise in the sea level," wrote Pope Francis, "can create extremely serious situations, if we consider that a quarter of the world's population lives on the coast or nearby, and that the majority of our megacities are situated in coastal areas" (Laudato Si 2015, p. 20).

Enhanced Tropical Storms

One of the most serious aspects of how climate change can alter weather patterns are the signs that it is leading to stronger tropical storms or (as they are often described) tropical cyclones. According to Dr. Wei Mei, a climate scientist at the Scripps Institution of Oceanography, in La Jolla, California, "Tropical cyclones are among the most devastating and destructive natural hazards on Earth" (Mei et al. 2015, p. 1). Since 2009, tropical cyclones have been divided into six categories, which are presented in Table 3.1.

The most serious tropical storms are typhoons or, as they are referred to in the western hemisphere, hurricanes. Typhoons develop in the northern hemisphere during the months of July to November in an area just north of the equator (Wisner et al. 2004). They develop when strong clusters of thunderstorms drift over ocean waters with a temperature of at least 26.5 °C. Warm air from these thunderstorms combines with warm air from the ocean's surface and begins rising; as this air rises there will be a reduction in air pressure on the surface of the ocean. As these clusters of thunderstorms

Table 3.1 The six categories of tropical storms

Type of storm	Wind speeds
Tropical depression	63 km/h or lower
Tropical storm	Between 63 and 89 km/h
Severe tropical storm	Between 90 and 119 km/h
Typhoon	Between 120 and 149 km/h
Severe typhoon	Between 150 and 190 km/h
Super typhoon	Greater than 190 km/h

Source: Abdullah et al. (2015)

consolidate into one large storm, trade winds blowing in opposite directions will cause the storm to begin spinning in a counterclockwise direction, while rising warm air causes air pressure to decrease at higher altitudes (Gonzalez 1994). Eventually, the storm will have a low pressure center (the eye) with no clouds and no winds, while winds in the outer part of the storm can be extremely strong (Gonzalez 1994).

All typhoons have four characteristics: low air pressure, strong winds, heavy rains, and storm surge (Bankoff 2003). The air pressure reduction associated with a typhoon can cause the sea level to rise by as much as 1 cm for every 1 mbar reduction in air pressure and there have been documented instances where sea levels have risen by 1.5 m due to air pressure reductions alone (Wang et al. 2005). This means that, at the same time that a typhoon (with its heavy rains and strong winds) comes ashore, the sea level will be higher due to the reduction of the atmospheric pressure (Gonzalez 1994). Although typhoons rapidly lose power as they move inland they are capable of causing massive amounts of damage to coastal areas, with a fully developed typhoon releasing the energy equivalent of an atomic bomb, and they move in an unpredictable manner becoming difficult to track (Wisner et al. 2004).

There is substantial scientific evidence indicating that climate change, by creating warmer sea surface temperatures, is leading to stronger tropical storms (Brecht et al. 2012; Combest-Friedman et al. 2012; Emanuel 2005, 2013; Emanuel et al. 2008; Knutson et al. 2010; Mei et al. 2015; Rozynski et al. 2009; Villarini et al. 2014; Webster et al. 2005). "Tropical cyclones," wrote Emanuel et al. (2008, p. 347), "next to floods are the leading cause of death and injury among natural disasters affecting developing countries." "As the climate changes during the 21st century," wrote Brecht et al. (2012, p. 120), "larger cyclonic storm surges and growing populations may collide in disasters of unprecedented size."

Climate Change: An Urgent Threat

Climate change poses an existential challenge to the human race. If it is not promptly addressed there will be, in the words of Pope Francis, "regional disasters which [will] eventually affect everyone" (Laudato Si 2015, p. 20). Having discussed the concept of climate change attention now turns to climate change and its impacts upon the Philippines.

References

Abdullah, K., Anukklarmphai, A., Kawasaki, T., & Neopmuceno, D. (2015). A tale of three cities: Water disaster policy responses in Bangkok, Kuala Lumpur, and Metro Manila. *Water Policy, 17*(S1), 89–113.

Bankoff, G. (2003). *Cultures of disaster: Society and natural hazard in the Philippines*. London: Routledge.

Bellard, C., Leclerc, C., & Courchamp, F. (2014). Impact of sea level rise on the 10 insular biodiversity hotspots. *Global Ecology and Biogeography, 23*(2), 203–212.

Brecht, H., Dasgupta, S., Laplante, B., Murray, S., & Wheeler, D. (2012). Sea-level rise and storm surges: High stakes for a small number of developing countries. *The Journal of Environment & Development, 21*(1), 120–138.

Chivers, D. (2011). *The no nonsense guide to climate change: The science, the solutions, the way forward*. Toronto: New Internationalist.

Combest-Friedman, C., Christie, P., & Miles, E. (2012). Household perceptions of coastal hazards and climate change in the Central Philippines. *Journal of Environmental Management, 112*(1), 137–148.

Crowley, T. J. (2000). Causes of climate change over the past 1000 years. *Science, 289*(5477), 270–277.

De Buys, W. (2011). *A great aridness: Climate change and the future of the American Southwest*. New York: Oxford University Press.

Ellwood, W. (2014). *The no nonsense guide to degrowth and sustainability*. Toronto: New Internationalist.

Emanuel, K. A. (2005). Increasing destructiveness of tropical cyclones over the past 30 years. *Nature, 436*(4), 686–688.

Emanuel, K. A. (2013). Downscaling CMIP5 climate models shows increased tropical cyclone activity over the 21st century. *Proceedings of the Natural Academy of Sciences, 110*(30), 12219–12224.

Emanuel, K. A., Sundararajan, R., & Williams, J. (2008). Hurricanes and global warming. *Bulletin of the American Meteorological Society, 89*(3), 347–367.

Gillett, N. P., Arora, V. K., Zickfield, K., Marshall, S. J., & Merryfield, W. J. (2011). Ongoing climate change following a complete cessation of carbon dioxide emissions. *Nature Geoscience, 4*(1), 83–87.

Gonzalez, E. B. (1994). Tropical cyclones and storm surges. R.S. Punongbayan, *Natural disaster mitigation in the Philippines: Proceedings, National Conference on Natural Disaster Mitigation* 19–21 October 1994, Quezon City, Philippines. Quezon City: Philippine Institute of Volcanology and Seismology: 11–8.

Haynes, K., & Tanner, T. M. (2015). Empowering young people and strengthening resilience: Youth-centered participatory video as a tool for climate change adaptation and disaster risk reduction. *Children's Geographies, 13*(3), 357–371.

Intergovernmental Panel on Climate Change. (2013). *Assessment report*. Retrieved from ipcc.ch

Knutson, T. R., McBride, J. L., Chan, J., Emanuel, K., Holland, G., Landsea, C. et al. (2010). Tropical cyclones and climate change. *Nature Geoscience, 3*(1), 157–163.

Laudato Si. (2015, May 24). *Encyclical Letter of the Holy Father Pope Francis*. Vatican City: Pope Francis.

Mei, W., Xie, S. P., Premeau, F., McWilliams, J. C., & Pasquero, C. (2015). Northwestern Pacific Typhoon intensity controlled by changes in ocean temperatures. *Science Advances, 4*(1), 1–8.

Min, S. K., Zhang, X., Zwiers, F. W., & Heger, G. C. (2011). Human contribution to more intense precipitation extremes. *Nature, 470*(7334), 378–526.

Overpeck, J. T., Otto-Bliesner, B. L., Miller, G. H., Muhs, D. R., Alley, R. B., & Kiehl, J. T. (2006). Paleoclimatic evidence for future ice-sheet instability and rapid sea-level rise. *Science, 311*(24), 1747–1750.

Overpeck, J. T., & Udall, B. (2010). Dry times ahead. *Science, 328*(5986), 1642–1643.

Romm, J. (2011). The next dust bowl. *Nature, 478*(7370), 450–451.

Rozynski, G., Hung, N. M., & Ostrowski, R. (2009). Climate change related rise of extreme typhoon power and duration over South-East Asia Seas. *Coastal Engineering Journal, 51*(3), 205–222.

Schneider, S. (2009). The worst-case scenario. *Nature, 458*(30), 1104–1105.

Thomas, V, Albert, J. R. G., & Perez, R. T. (2013). *Climate-related disasters in Asia and the Pacific. Asian Development Bank Economics.* Working paper series no. 358. Manila: Asian Development Bank.

Trenberth, K., Overpeck, J., & Solomon, S. (2004). Exploring drought and its implications for the future. *Eos, 85*(3), 26–28.

Villarini, G., Lavers, D. A., Scoccimarro, E., Zhao, M., Wehner, M. F., Vecchi, G. A., et al. (2014). Sensitivity of tropical cyclone rainfall to idealized global-scale forcings. *Journal of Climate, 27*(12), 4622–4641.

Villarini, G., Smith, J. A., & Vecchi, G. A. (2013). Changing frequency of heavy rainfall over the Central United States. *Journal of Climate, 26*(1), 351–357.

Walch, C. (2014). Collaboration or obstruction? Rebel group behavior during natural disaster relief in the Philippines. *Political Geography, 43*, 40–50.

Wang, Y. H., Lee, I. H., & Wang, D. P. (2005). Typhoon induced extreme coastal surge: A case study at Northeast Taiwan in 1994. *Journal of Coastal Research, 21*(3), 548–552.

Webster, P. J., Holland, G. J., Curry, J. A., & Chang, H. R. (2005). Changes in tropical cyclone number, duration, and intensity in a warming environment. *Science, 309*(5742), 1844–1846.

Westra, S., Alexander, L. V., & Zwiers, F. W. (2013). Global increasing trends in annual maximum daily precipitation. *Journal of Climate, 26*(3), 3904–3918.

Wisner, B., Blaikie, P., Cannon, T., & Davis, I. (2004). *At risk: Natural hazards, people's vulnerability and disasters* (2nd ed.). London: Routledge.

Yang, L., Smith, J. A., Wright, D. B., Baeck, M. L., Villarini, G., Tian, F., et al. (2013). Urbanization and climate change: An examination of nonstationarities in urban flooding. *Journal of Hydrometeorology, 14*(6), 1791–1809.

Chapter 4
An Archipelago of Hazards

The Philippines is one of the world's most hazard prone countries. Located on the Pacific Ring of Fire, the archipelago is vulnerable to geophysical hazards, such as earthquakes, tsunamis, and volcanoes (Bankoff 2003; Holden and Jacobson 2012) as well as climatological hazards such as El Niño induced drought (Holden 2013) and tropical cyclones (Holden 2015). Arguably the most salient of these hazards, particularly insofar as climate change is concerned, are typhoons. One of the strongest tropical cyclones ever recorded in the world, typhoon Haiyan, locally referred to as Yolanda, swept widespread havoc and destruction across the Philippines on November 8, 2013. According to the United Nations' Office of Coordinated Humanitarian Affairs, 14 million people in the Philippines were affected by this typhoon, including thousands of precious lost human lives and 4 million people were displaced, including 1.8 million children. Nearly, 12,000 babies were born in the month following the typhoon.

The Philippines are highly vulnerable to typhoons and are situated where more typhoons are formed than anywhere else in the world (Department of Environment and Natural Resources Climate Change Office 2010; Thomas et al. 2013). Each year the archipelago is exposed to roughly 25% of the total number of such events worldwide (Yumul et al. 2001). These typhoons originate south and east of the archipelago during the months of July to November and travel in a northwesterly direction mainly affecting the eastern half of the country. The most heavily affected portions of the Philippines are Samar, the Bicol Region of southeast Luzon, and northern Luzon (Bankoff 2003). "Eastern Samar," wrote Haynes and Tanner (2015, p. 361), "is one of the most disaster affected areas in the country, exposed to high rainfall and frequent typhoons." Bicol's frequent exposure to typhoons makes it one of the poorest regions of the Philippines (Atienza 2015). "Metro Manila," wrote Abdullah et al. (2015, p. 92), "sits astride the typhoon belt and experiences heavy torrential rains during the typhoon season."

Examining the flood vulnerability and adaptation of urban poor communities in Metro Manila, Emma Porio (2011) found that they are more prone to experiencing flooding and inundation when typhoons occur. The effects of increasingly stronger

© The Author(s) 2017
W. Holden et al., *Ecological Liberation Theology*, SpringerBriefs in Geography,
DOI 10.1007/978-3-319-50782-8_4

typhoons hitting the Philippines, poses a great challenge to local coastal and riverine communities' ecological-environmental systems and the wellbeing and security of residents. Along river shores and bay areas, where large numbers of poor families reside, land and infrastructural development initiatives of the government and the private sector often collide with the pursuit of security and wellbeing of local residents. This gives rise to particular *place-based vulnerabilities* and *sector-specific* patterns of response and adaptation to flooding and environmental degradation. The interaction of place-based and sector-specific vulnerabilities, especially among the urban poor in informal settlements, heightens and compromises both the environmental and human security needs of the people and their communities. The vulnerability levels of families or households living in these environmentally vulnerable places then becomes heightened because their capacity to recover from flood impacts is also compromised by their poverty/low incomes, and fragile occupational or livelihood bases. The dangers typhoons pose to these communities around the archipelago will be amplified by climate change and attention now turns to this amplification.

Sea Level Rise

As an archipelago with 36,289 km of coastline, the Philippines is particularly susceptible to sea level rise (Kehew et al. 2013). The Philippines is ranked as the third most vulnerable country in the world to natural hazards and climate change (see Table 4.1) and the majority of all food produced in the archipelago is produced on land barely above sea level (Broad and Cavanagh 2011). The sea level in the waters surrounding the Philippines are at historic highs and, since 1970, sea level measured at Legaspi, in Bicol, has risen by more than 200 mm (Lander et al. 2014). Over time, the archipelago can expect to lose between 7% and 17% of its islands due to sea level rise and the Philippines, along with the Caribbean and Sundaland, is one of the three places in the world predicted to lose the most land due to climate change

Table 4.1 Vulnerability of countries to natural hazards and climate change

Vulnerability ranking	Country
1	Vanuatu
2	Tonga
3	**Philippines**
4	Guatemala
5	Bangladesh
6	Solomon Islands
7	Costa Rica
8	Cambodia
9	Timor-Leste
10	El Salvador

Source: Alliance Development Works (2012)

(Bellard et al. 2014). Higher sea levels exacerbate the seriousness of typhoons because, with higher sea levels, a typhoon's storm surge will be even more intense and dangerous even if there are no changes in the intensity of the typhoons themselves (Knutson et al. 2010).

Typhoons Carrying More Moisture

The typhoons impacting the Philippines, even those of low intensity, are becoming accompanied by heavier rainfall events (Thomas et al. 2013). The run off from these heavy rainfall events is extremely difficult to model and predict and severe flooding can occur (Clutario and David 2014). Consider for example Typhoon Ketsana (referred to in the Philippines as Typhoon Ondoy), which impacted Manila on 26 September 2009 depositing over 340 mm of rain fell in only 6 h (Holden and Jacobson 2012). This typhoon shattered the previous Philippine precipitation record set 42 years earlier in 1967 when 340 mm fell over a 24 h period. Amalie Obusan, a climate and energy campaigner for Greenpeace Southeast Asia, has no doubt that Typhoon Ketsana and Typhoon Parma (referred to in the Philippines as Typhoon Pepeng), which also impacted the archipelago during September 2009, were the result of climate change due to the amounts of rain that fell relative to historical data (Obusan 2009). Obusan recalled how after Typhoon Ketsana, the Philippine Red Cross asked Greenpeace Southeast Asia to use its inflatable boats (usually used for campaign purposes) to rescue people from the flooded streets of Marikina City in Metro Manila; "these typhoons were unlike anything ever seen in Philippine history" (Obusan 2009, interview). Carlos Conde, a journalist who has written about climate change for the International Herald Tribune, clearly attributes these typhoons to climate change as the volume of rain was simply incredible (Conde 2009). Conde finds it difficult not to attribute such powerful typhoons to climate change and views the combination of the archipelago's vulnerability to typhoons with climate change as a situation where "there are bound to be problems" (Conde 2009, interview).

Unpredictable Typhoons

Tropical storm tracks are very hard to predict on sub seasonal and seasonal timescales (Mei and Xie 2015), although the track of a particular tropical storm can be predicted with some confidence (Mei 2015). In Dr. Mei's opinion, we are not sure that climate change is causing the typhoons that impact the Philippines to track in an *east* to *west* trajectory instead of their normal *southeast* to *northwest* trajectory (Mei 2015). Nevertheless, despite the absence of scientific certainty regarding the impact of climate change on tropical storm tracks, typhoons in the Philippines seem to have become more unpredictable. In December 2011, Tropical Storm Sendong tracked from east to west over Mindanao, causing severe flooding in the city of Cagayan de Oro and killing one thousand people; in December 2012, Typhoon Pablo tracked from east to west over Mindanao killing a similar number of people and forcing one million people to evacuate their homes (Internal Displacement Monitoring Center 2013).

Stronger Typhoons

On 8 November 2013 Super Typhoon Haiyan (known in the Philippines as Super Typhoon Yolanda) ravaged the Philippines causing more than 6200 deaths with 1785 people being reported missing (Mei et al. 2015) and more than four million people being displaced (Yamada and Galat 2014). More than 2 months after Haiyan passed through the archipelago bodies were still being found and almost ten billion dollars' worth of damage was inflicted upon the Philippine economy, approximately 93% of which was not covered by any form of insurance (Abdullah et al. 2015). Haiyan was the strongest tropical cyclone to ever make landfall in the world with wind speeds estimated at between 305 and 314 km/h (Yamada and Galat 2014) and it had a storm surge as high as 7.5 m above sea level (Lander et al. 2014). In Tacloban, Leyte, where Haiyan caused its most destructive damage, much of the land are is less than 5 m above sea level and Haiyan had "nearly the same force and rapidity of a destructive tsunami" (Lander et al. 2014, p. S.114). Dr. Mei stated he was "shocked" by the strength of Haiyan (Mei 2015, interview).

Was Super Typhoon Haiyan the result of climate change or was it merely an aberration? Takayabu et al. (2015) conducted statistical simulations of Haiyan's intensity using general circulation models and these showed that Haiyan was the result of radiative forcing due to anthropogenic climate change. Research conducted by Dr. Mei and his colleagues on Pacific typhoons found that the intensity of typhoons over the last 10 years has been, on average, the strongest of the last 60 years (Mei et al. 2015). In Dr. Mei's opinion this strengthening of tropical cyclones might be in part due to climate change and the continued ocean warming under moderate estimates of climate change may cause a further 14% increase in typhoon intensity by the year 2100 (Mei 2015; Mei et al. 2015). This "strengthened typhoon intensity poses heightened threats to human society" (Mei et al. 2015, p. 4). In the future there will be a shift toward stronger typhoons especially in terms of mean intensity, although the strength of individual storms cannot be predicted (Mei 2015). In Dr. Mei's opinion, people in the Philippines have to be concerned about the intensity of tropical cyclones in the coming future; the water east of the Philippines is getting warmer and this may well cause stronger typhoons (Mei 2015). Dr. Mei is not sure what people in the archipelago can do to reduce their vulnerability to tropical cyclones; if he lived in the Philippines he would be very worried about tropical cyclones (Mei 2015).

Synergistic Relations Between Typhoons and Other Natural Hazards

One of the most dangerous dimensions of typhoons, enhanced in their potency by climate change, are the synergistic relations typhoons have with other types of anthropogenic environmental degradation in the Philippines, most notably deforestation and land subsidence. Much of the local deforestation is attributable to illegal logging (Van der Ploeg et al. 2011) and its effect is most notable with respect to the diminution of old-growth rain forest, which covered approximately 70% of the national land area in 1900 but had fallen to only 8% of the national land area in 1992 (Heaney and Regalado 1998). Should there be a typhoon (with its associated heavy rains) landslides will occur much more quickly on deforested hillsides (Alliance Development Works 2012). Deforestation also reduces the mitigating effect availed to inland locations by their remoteness from the ocean since the deforestation of coastal areas allows typhoons to penetrate further inland and inflict damage over wider areas (Myers 1988). Groundwater withdrawal, necessitated by an El Niño induced drought, can lead to land subsidence (Rodolfo and Siringan 2006). As water is progressively removed from an aquifer, the density of that aquifer diminishes and, over time, the level of the ground will be reduced. Combine land subsidence with higher sea levels brought on by the reduction of air pressure during a typhoon, the higher sea levels brought on by climate change, and the vulnerability of coastal areas to typhoons increases even more and such areas become exceptionally susceptible to flooding brought on by the typhoon's storm surge. This generates a perverse situation where a lack of rain (a drought) leads to ameliorative measures (groundwater withdrawal) that increases vulnerability to the problems occurring when the rains return during typhoon season.

Climate Change and Enhanced Geophysical Hazards

Ultimately, it is quite conceivable that anthropogenic climate change will lead to enhanced geophysical hazards in the Philippines, such as earthquakes, tsunamis and volcanism. It is well established that climate change will lead to a reduction of the thickness of the Antarctic and Greenland ice sheets (Bellard et al. 2014; Overpeck et al. 2006). As the thickness of these ice sheets are reduced there will be a reduction of the weight being borne by the tectonic plates that rest upon the Earth's outer mantle and the corresponding isostatic adjustment of the outer mantle could generate more tectonic activity thus more seismic and volcanic activity (Brandes et al. 2011; Galgana and Hamburger 2010; Grollimund and Zoback 2001; Poutanen and Ivins 2010). Such effects may take several hundred years to fully manifest themselves but eventually the warming of the Earth's atmosphere could well lead to more geophysical hazards as well as meteorological hazards.

References

Abdullah, K., Anukklarmphai, A., Kawasaki, T., & Neopmuceno, D. (2015). A tale of three cities: Water disaster policy responses in Bangkok, Kuala Lumpur, and Metro Manila. *Water Policy, 17*(S1), 89–113.

Alliance Development Works. (2012). *World risk report 2012.* Berlin: Alliance Development Works.

Atienza, M. E. L. (2015). People's views about human security in five Philippine municipalities. *Disaster Prevention and Management, 24*(4), 448–467.

Bankoff, G. (2003). *Cultures of disaster: Society and natural hazard in the Philippines.* London: Routledge.

Bellard, C., Leclerc, C., & Courchamp, F. (2014). Impact of sea level rise on the 10 insular biodiversity hotspots. *Global Ecology and Biogeography, 23*(2), 203–212.

Brandes, C., Polom, U., & Winsemann, J. (2011). Reactivation of basement faults: Interplay of ice-sheet advance, glacial lake formation and sediment loading. *Basin Research, 23*(1), 53–64.

Broad, R., & Cavanagh, J. (2011). Reframing development in the age of vulnerability: From case studies of the Philippines and Trinidad to new measures of rootedness. *Third World Quarterly, 32*(6), 1127–1145.

Clutario, M. V. A., & David, C. P. C. (2014). Event-based soil erosion estimation in a tropical watershed. *International Journal of Forest, Soil, and Erosion, 4*(2), 51–57.

Conde, C. H. (2009, November 12). Journalist. *International Herald Tribune.* Personal interview conducted by William Holden, Quezon City, Philippines.

Department of Environmental and Natural Resources Climate Change Office (DENR). (2010). *The Philippine strategy on climate change adaptation.* Quezon City: Department of Environmental and Natural Resources Climate Change Office.

Galgana, G. A., & Hamburger, M. W. (2010). Geodetic observations of active intraplate crustal deformation in the Wabash Valley seismic zone and the southern Illinois Basin. *Seismological Research Letters, 81*(5), 699–714.

Grollimund, B., & Zoback, M. D. (2001). Did deglaciation trigger intraplate seismicity in the New Madrid Seismic zone? *Geology, 29*(2), 175–178.

Haynes, K., & Tanner, T. M. (2015). Empowering young people and strengthening resilience: Youth-centered participatory video as a tool for climate change adaptation and disaster risk reduction. *Children's Geographies, 13*(3), 357–371.

Heaney, L. R., & Regalado, J. C. (1998). *Vanishing treasures of the Philippine rain forest.* Chicago: The Field Museum.

Holden, W. N. (2013). Neoliberal mining amid El Nino induced drought in the Philippines. *Journal of Geography and Geology, 5*(1), 58–77.

Holden, W. N. (2015). Mining amid typhoons: Large-scale mining and typhoon vulnerability in the Philippines. *The Extractive Industries and Society, 2*(3), 445–461.

Holden, W. N., & Jacobson, R. D. (2012). *Mining and natural hazard vulnerability in the Philippines: Digging to development or digging to disaster?* London: Anthem Press.

Internal Displacement Monitoring Center. (2013). *Living in the shadows: Displaced lumads locked in a cycle of poverty.* Geneva: Internal Displacement Monitoring Center.

Kehew, R. B., Kolisa, M., Rollo, C., Callejas, A., Alber, G., & Ricci, L. (2013). Formulating and implementing climate change laws and policies in the Philippines, Mexico (Chiapas), and South Africa: A local government perspective. *Local Environment, 18*(6), 723–737.

Knutson, T. R., McBride, J. L., Chan, J., Emanuel, K., Holland, G., Landsea, C. et al. (2010). Tropical cyclones and climate change. *Nature Geoscience, 3*(1), 157–163.

Lander, M., Guard, C., & Camargo, S. J. (2014). Super Typhoon Haiyan [in "State of the Climate in 2013"]. *Bulletin of the American Meteorological Society, 95*(7), S112–S114.

Mei, W. (2015, November 4). Climate Scientist. *Scripps Institution Of Oceanography.* Personal interview conducted by William Holden, La Jolla, California.

Mei, W., & Xie, S. P. (2015). Forced and internal variability of tropical cyclone track density in the western north Pacific. *Journal of Climate, 28*(1), 143–167.

Mei, W., Xie, S. P., Premeau, F., McWilliams, J. C., & Pasquero, C. (2015). Northwestern Pacific typhoon intensity controlled by changes in ocean temperatures. *Science Advances, 4*(1), 1–8.

Myers, N. (1988). Environmental degradation and some economic consequences in the Philippines. *Environmental Conservation, 15*(3), 303–311.

Obusan, A.C. (2009, November 4). Climate and Energy Campaigner. *Greenpeace Southeast Asia.* Personal interview conducted by William Holden, Quezon City, Philippines.

Overpeck, J. T., Otto-Bliesner, B. L., Miller, G. H., Muhs, D. R., Alley, R. B., & Kiehl, J. T. (2006). Paleoclimatic evidence for future ice-sheet instability and rapid sea-level rise. *Science, 311*(24), 1747–1750.

Porio, E. (2011). Vulnerability, adaptation and resilience among marginal riverine communities in Metro Manila. *Asian Journal of Social Science, 45*(2011), 425–445.

Poutanen, M., & Ivins, E. R. (2010). Upper mantle dynamics and quaternary climate in cratonic areas (DynaQlim)-understanding the glacial isostatic adjustment. *Journal of Geodynamics, 50*(1), 2–7.

Rodolfo, K. S., & Siringan, F. S. (2006). Global sea-level rise is recognized, but flooding from anthropogenic land subsidence is ignored around Northern Manila Bay, Philippines. *Disasters, 30*(1), 118–139.

Takayabu, I., Hibino, K., Sasaki, H., Shiogama, H., Mori, N., Shibutani, Y., et al. (2015). Climate change effects on the worst-case storm surge: A case study of typhoon Haiyan. *Environmental Research Letters, 10*(6), 1–9.

Thomas, V., Albert, J. R. G., & Perez, R. T. (2013). *Climate-related disasters in Asia and the Pacific. Asian Development Bank Economics.* Working paper series no. 358. Manila: Asian Development Bank.

Van der Ploeg, J., van Weerd, M., Masipiquena, A. B., & Persoon, G. A. (2011). Illegal logging in the Northern Sierra Madre Natural Park, the Philippines. *Conservation and Society, 9*(3), 202–215.

Yamada, S., & Galat, A. (2014). Typhoon Yolanda/Haiyan and climate justice. *Disaster Medicine and Public Health Preparedness, 8*(5), 432–435.

Yumul, G. P., Cruz, N. A., Servando, N. T., & Dimalanta, C. B. (2001). Extreme weather events and related disasters in the Philippines, 2004–08: A sign of what climate change will mean? *Disasters, 35*(2), 362–382.

Chapter 5
Neoliberalism Exasperates the Problem of Climate Change

Since the mid-1970s an aggressive new strain of capitalism has come to dominate the world and it is called "neoliberalism." The word "liberalism" is used because neoliberalism looks back to the classical economists, such as Adam Smith, for its inspiration but it is an aggressive new variant of liberalism lacking the empathy for other humans that the classical economists articulated; consequently, the prefix "neo" is added to the word "liberalism." A widely used definition of neoliberalism is that used by Harvey (2006, p. 2) who defined it as:

> A theory of political economic practices that proposes that human well-being can best be advanced by liberating individual entrepreneurial freedoms and skills within an institutional framework characterized by strong private property rights, free markets, and free trade.

A crucial dimension of neoliberalism is a wide-ranging reduction of state involvement in the economy (Harvey 2011). The state is viewed by neoliberals as egregiously inefficient and, wherever possible, state operated enterprises must be privatized. Corporations are viewed as infallible and all problems have market-based solutions (Mirowski 2013). All Policy prescriptions involving government regulation are anathema to neoliberalism. Instead, neoliberalism rather attempts to harness and reign in the state for its own onward expansion.

Some 25 years ago, Carol Smith (1996, p. 25) argued that the ideological underpinnings of development must be separated out and examined in conjunction with the role of the state and in relation to international organizations and the global interstate system. She discussed and analysed the development of the world capitalist model that took off, after World War II (1945-mid-1970s). International development, after WWII, was mainly capital intensive; money was poured into Third World states (now commonly referred to as "the global south") to build basic infrastructure such as roads and railroads and small and large-scale capitalist farming techniques in agriculture and fisheries. Such investment was designed to increase the speed and quantity of production and get products to the market. Local transportation systems were improved mainly to move the products out through the

© The Author(s) 2017
W. Holden et al., *Ecological Liberation Theology*, SpringerBriefs in Geography,
DOI 10.1007/978-3-319-50782-8_5

ports. Though not explicitly publicized, these investments were primarily intended to benefit the material conditions of the donor countries. The result of this huge influx of capital was that Third World states became incredibly indebted (Stutz and Warf 2008). Consequently and, approximately, coinciding with the rise of Reaganomics in the United States and, then, the New World Order following the dissolution of the Soviet Union, which ended the Cold War in 1991, the response of big donor organizations has been focused mainly on restructuring Third World debt by lowering interest rates and stretching out debt payments. This strategy earlier developed, in 1987, by the World Bank and International Monetary Fund to bail out the Philippine economy left bankrupt by the fall of the Marcos regime had a built in caveat: in exchange for stretching out the Philippine government's loan payments, it contractually agreed to further liberalize the economy by selling governmentally owned public utility services and public health facilities, competitively funding non-government sponsored education programs, rather than, as before in the post-WWII period, channelling it through the government's Department of Education in charge of the public school system, to further so-called economic development (Holden and Nadeau 2010, p. 92; Smith 1996, p. 25). This meant that they wanted the Philippine government to create the conditions necessary for attracting business entrepreneurs, for profit non-government organizations, and multinational corporations and to do this, the government was required, by the lending organizations, to also open its markets, keep wages low, and offer other kinds of incentives to international businesses (Tujan 2007). That is, in exchange for, and as a condition of receiving a loan or the restructuring of an existing loan that still had to be paid back, big lender organizations were allowed to intervene in planning and managing the Philippine economy (Stutz and Warf 2008); which was the same strategy used to bailout other affected countries as Thailand and South Korea, whose economies capsized during the 1999 Asian Economic Crisis.

Past and present capitalist globalization and development approaches geared toward improving the economy are based on an obsolete and dysfunctional yet still on-going myth of "catching up" development or, to use another widely known expression, "trickle-down" development. The myth of catching up development is based on the false premise that the most appropriate and best model of the affluent society is that of better off countries (formerly, Western Europe; the United States; Canada; Japan; Singapore; Taiwan) that underwent rapid modernization and industrialization in the post-WWII period of the last twentieth century. This effectively meant that poor countries, which earlier elected to take the same path to industrialization, as the so-called modern industrial and better off societies would eventually achieve the same level of development. However, as one can see in the United States (a supposedly rich country), its poor neighbourhoods have become a ubiquitous feature of the landscape and are terrains of danger and violence, where drugs, guns, and gangs reign. Many nominally major U.S. corporations and businesses, rather than investing locally are now outsourcing, relocating and transnationalizing to avoid paying taxes, and hiring people abroad at lower wages for sake of making ever bigger profit. There are many working and unemployed people struggling in near abject poverty in the United States. In contemporary China and India, the new economic superpowers, the local majorities, on a much larger scale, are lower middle class and poor, living in conditions worsened by lack of funding for public education and social services and widespread pollution of the air, land, and waters as a result of so-called globalization and development. At the same time, the questionable affluence of the developed world exists simultaneously with a decline in the material conditions of the majority of the contemporary world's poorest communities (especially in impoverished areas, where the rich are super rich and the majority of ordinary people are poor, like the Philippines, internally ravished by war, human rights violations, natural disasters, and other disasters coming from human manufactured hazards).

Neoliberalism has become a hegemonic discourse in the new millennium, something that has achieved the status of being taken for granted or, more than that, as Peet (2003, p. 4) stated, has "achieved the supreme power of being widely taken as scientific and resulting in an optimal world." Worldwide, neoliberal doctrines and right-wing political parties have achieved a pervasive dominance (Mirowski 2013). Acts of protest against neoliberalism are taken "as an offence against reason, progress, order, and the best world ever known" (Peet 2003, p. 4).

Inherent in the neoliberal reluctance to provide an institutional response to climate change is neoliberalism's tendency to deny climate change. "Science suggests that human action is contributing to global warming," wrote Harvey (2011, p. 187), "reducing opponents (usually funded by the energy lobby) to the astonishing claim that global warming is a hoax perpetrated by scientists upon the world's population." At the heart of anthropogenic climate change are greenhouse gas emissions. Widespread government efforts to dramatically reduce greenhouse gas emissions are, in the minds of neoliberals, a Trojan horse concealing a socialist takeover of the economy. Consequently, neoliberalism has engaged in a three-staged campaign to act against actions to meaningfully reduce greenhouse gas emissions (Mirowski 2013).

The first step is to deny the scientific evidence of climate change. Through corporate capitalism's control of the media (outlets such as Rupert Murdoch's Fox News come readily to mind) the few fringe "scientists" that deny climate change are accorded equal air time to mainstream climatologists, and this generates the (incorrect) impression in the minds of members of the public that there is a vigorous scientific debate about climate change. Mirowski (2013, p. 226) refers to this as "agnotology" and it is "the intentional manufacture of doubt and uncertainty in the general populace for specific political motives." This first strategy has been highly effective; according to Mirowski (2013, p. 243):

> As the climate crisis worsens, the general populace appears to recoil from the worsening news, fleeing the facts, preferring to disdain the science rather than confront the crisis. In one instance, the Harris polling organization claimed that in 2007, 71 percent of all Americans said they believed that continued combustion of fossil fuels resulted in climate change, whereas by June 2011 it had dropped to 41 percent.

The second step is to *pretend* to act to reduce greenhouse gas emissions by developing a market for greenhouse gas emissions, the so-called "cap and trade" solution. This solution makes it *appear* that something is being done to address the problem of climate change while, in reality, no meaningful reductions in greenhouse gas are imposed upon corporate capitalism that may impinge the profitability of it's using the Earth's atmosphere as an open sewer. As Mirowski (2013, p. 338) wrote:

> The project to institute markets in pollution permits is a neoliberal mid-range strategy, better attuned to appeal to neoliberal governments, NGOs, and the more educated segments of the populace, not to mention the all-important [finance, insurance, and real estate] sector of the economy. In effect, this strategy is an elaborate bait-and-switch, where political actors originally bent upon using state power to curb emissions are instead diverted into the endless technicalities of the institution and maintenance of novel markets for carbon permits.

The third step is geoengineering. Instead of acting to reduce greenhouse gas emissions actions are undertaken to reduce their environmental effects. The classic examples of geoengineering are carbon sequestration and releasing aerosols into the atmosphere to offset the retention of solar energy. These solutions are completely untried and untested and they act as "end of pipe" solutions as opposed to "ending the pipe." They are akin to a heavy smoker spending money on cough drops and asthma medicine, while continuing to chain smoke cigarette after cigarette after cigarette after cigarette. In the words of Mirowski (2013, p. 340):

> The final neoliberal fallback is geoengineering, which derives from the core neoliberal doctrine that entrepreneurs, unleashed to exploit acts of creative destruction, will eventually innovate market solutions to address dire economic problems. This is the whiz-bang futuristic science fiction side of neoliberalism, which appeals to male adolescents and Silicon Valley entrepreneurs almost as much as do the novels of Ayn Rand.

These three steps act in close coordination with each other: denial postpones addressing the issue of climate change by creating doubt about whether or not it is even a problem; carbon permit trading creates the impression that something is being done, when in fact there are no actual reductions in emissions; and geoengineering merely treats the symptoms of global warming without curtailing emissions (Mirowski 2013).

References

Harvey, D. (2006). Neo-liberalism as creative destruction. *Geografiska Annaler: Series B: Human Geography, 88*(2), 145–158.

Harvey, D. (2011). *The enigma of capital and the crises of capitalism.* Oxford: Oxford University Press.

Holden, W., & Nadeau, K. (2010). Philippine liberation theology and social development in anthropological perspective. *Philippine Quarterly of Culture and Society, 38*(2), 89–129.

Mirowski, P. (2013). *Never let a serious crisis go to waste: How neoliberalism survived the financial meltdown.* London: Verso.

Peet, R. (2003). *Unholy trinity: The IMF, World Bank and WTO.* London: Zed Books.

Smith, C. (1996). Development and the state: Issues for anthropologists. In *Transforming societies, transforming anthropology*. Ann Arbor: University of Michigan Press.

Stutz, F. P., & Warf, B. (2008). *The world economy: Resources, location, trade, and development* (5th ed.). Toronto: Pearson Education.

Tujan, A. A. (Ed.). (2007). *Jobs and justice: Globalization, labor rights, and workers' resistance.* Quezon City: IBON Books.

Chapter 6
Neoliberalism in the Philippines

Neoliberalism's policy prescription is an almost exclusive reliance upon the market as the institution to be used for resource allocation and the goal of neoliberalism is to increase foreign investment (McCarthy 2007). Since the state is presumed to be inefficient, neoliberalism eschews any role for the state in responding to the needs of the populace (McCarthy 2007). For example, wages can be kept inordinately low to lure outside corporations and businesses to invest in the Philippines. Under the auspices of neoliberalism, major multilateral agencies such as the Asian Development Bank, World Bank, International Monetary Fund, and World Trade Organization, 'have become increasingly aggressive in their willingness to look inside countries, evaluate their governance structures, and recommend both sweeping and highly specific changes' (McCarthy 2007, p. 40). These multilateral agencies have called for foreign investors in developing countries to be guaranteed parity rights and protections against expropriation, and to be allowed to freely move investment funds and profits into, and out of, a country as they wish (McCarthy 2007).

In the Philippines, there is a concentration of wealth and power in the hands of a powerful oligarchy, while the majority of the population lives in poverty (Hutchcroft 1998). Raluto (2015, p. 4) goes so far as to argue that the relationship between local poverty and oppression is the cause of the local environmental crisis. Tyner (2009, p. 2) described the archipelago as a county of "haves and have-nots" and "a county of the super-rich and the abject poor." A pithy description of the contemporary Philippines is that provided by Mason (2012, p. 201):

> If your vision of capitalism is one in which a genetically predestined elite runs everything, where democracy is a vibrant sham, where the minds of the poor are controlled by religion, TV and lotteries, and where patronage and graft is rife, then the Philippines is the ideal embodiment of it.

The response of successive governments to the poverty of the archipelago's inhabitants has been an adherence to the policy agenda of neoliberalism (Bello et al. 2009. "The Philippines," wrote Broad and Cavanagh (2011, p. 1134), "has long been a 'poster child' of an open economy." The government has eschewed

© The Author(s) 2017
W. Holden et al., *Ecological Liberation Theology*, SpringerBriefs in Geography,
DOI 10.1007/978-3-319-50782-8_6

intervening in the economy and has decided that the best course for development is what Tadem (2011, p. 8) described as "an economy best left to the market with a minimum of government intervention." A succinct description of neoliberalism in the Philippines is that provided by Quimpo (2008, p. 49):

> Under pressure from the International Monetary Fund and World Bank, the government has pursued various 'structural adjustment programs' since the Marcos era, stressing first trade liberalization, then debt repayment, and finally free-market transformation marked by rapid deregulation, privatization, and trade and investment liberalization.

References

Bello, W., Docena, H., de Guzman, M., & Malig, M. L. (2009). *The anti-development state: The political economy of permanent crisis in the Philippines*. Manila: Anvil Publishing.

Broad, R., & Cavanagh, J. (2011). Reframing development in the age of vulnerability: From case studies of the Philippines and Trinidad to new measures of rootedness. *Third World Quarterly, 32*(6), 1127–1145.

Hutchcroft, P. (1998). *Booty capitalism in the Philippines*. Ithaca: Cornell University Press.

Mason, P. (2012). *Why it's kicking off everywhere: The new global revolutions*. London: Verso.

McCarthy, J. (2007). Privatizing conditions of production: Trade agreements as neoliberal environmental governance. In N. Heynen, J. Mccarthy, S. Prudham, & P. Robbins (Eds.), *Neoliberal environments: False promises and unnatural consequences* (pp. 38–50). London: Routledge.

Quimpo, N. G. (2008). *Contested democracy and the left in the Philippines after Marcos*. New Haven: Yale University Press.

Raluto, R. D. (2015). *Poverty and ecology at the crossroads, towards an ecological theology of liberation in the Philippine context*. Quezon City: Ateneo de Manila University Press.

Tadem, T. S. E. (2011). Introduction: Examining global civil society movements in the Philippines. In T. S. E. Tadem (Ed.), *Global civil society movements in the Philippines* (pp. 1–24). Manila: Anvil Publishing.

Tyner, J. A. (2009). *The Philippines: Mobilities, identities, globalization*. London: Routledge.

Chapter 7
Alternative Development Approach of Ecological Liberation Theology

Neoliberal approaches to solving problems of climate change are further increasing the gap between the rich and poor, by promoting development projects in the affected communities that may empower some people but not everyone. Some individuals benefit by being given jobs and material goods but other individuals are left wanting. A creative alternative development approach to that of neoliberal capitalism and globalization is being put into practice by some of the frontline churches and community organizers partnering with Philippine faith-based communities to build what Armstrong (2008) refers to as communities of compassion, also called social geographies of compassion, from the perspective of ecological liberation theology. This ecological theology movement is a revolutionary social and environmental peace and justice movement that bridges differences and brings people together for a common cause. Theoretically, it articulates not only reading scripture in accordance with the spirit of the letter, across different world religions and philosophies but, also, a creative and non-dogmatic eco-feminist and neo-Marxist perspective. However, while Marx argued that religion serves to legitimate the privileges of elite classes by disguising socio-cultural and economic inequities of production, this brief takes the more flexible position that religion can liberate as well as oppress.

The concept of religion used in eco-liberation theology is not some abstract or universal notion, but, rather is discursively, practically, materially, and specifically, grounded (Nadeau 2002, p. 75). Marx (1972, p. 14) looked at religion as an aspect of ideology. He showed how humans think about their practices and circumstances and, then, how thoughts spur action. In this sense, Marx used religion as an aspect of ideology to study the content and structure of the ruling classes. He showed how these classes used ideology in such a way to persuade other classes to provide them services on a voluntary or abject and low wage basis.

Marx's concepts, like any other theoretical ideas, are best seen as entry points for social analysis. They offer an open-ended approach, not a fixed model, for the study of social and religious movements from an insider-outsider perspective. Late twentieth century Philippine liberation theologians, like De la Torre (1986), Gaspar (1990), and Cacayan and Miclat (1991) wrote about the local organized Basic

© The Author(s) 2017
W. Holden et al., *Ecological Liberation Theology*, SpringerBriefs in Geography,
DOI 10.1007/978-3-319-50782-8_7

Christian Community movement as a self-conscious and enlightened new social movement that acted on its own behalf. In the 1990s, Nadeau (1995, p. 31; see, also, 2002) observed that the social contexts in which Basic Christian Communities interacted with other people (employers, landlords, military personnel, government officials, disinterested neighbors) and the institutional Church were situations in which new ideas and cultural forms were continuously being introduced, negotiated, and transformed. Members were not necessarily subordinated to the Church hierarchy, but utilized resources and opportunities the Church provided to promote their own interests. Basic Christian Community participants consciously employ traditional ideas to resist being fragmented by capitalist relations of production as when farmers collaborate together in planting and harvesting their crops or building each others' homes, not for wages but in the spirit of community. Their struggle for liberation continues into the present day, to be expressed through the agencies of Christian ceremonial activities and symbols (Ileto 1979; Gasper, pers. comm. 1991). The nationwide Basic Christian Community movement was organized, in full force, in the context of the Martial Law regime of the Ferdinand Marcos dictatorship, from 1972 until 1986. At that time, except for the church, most major institutions—the congress, courts, political parties, labor organizations, newspapers, and public broadcasting networks—were severely repressed by Marcos's military. As a result, the political significance of the Church grew clearer. It became the spokesperson for the rights of the poor and oppressed. Church leaders referred to the social teachings of Vatican II (1962–1965) to promote social justice (Bolasco and Yu 1981; Fabros 1988, p. 14; Youngblood 1990, Ch. 4).

During the Marcos dictatorship, the government lumped all Christians critical of martial law into a radical left (Youngblood 1990). Press and media campaigns were directed against the Basic Christian Communities, bishops sympathetic to martial law were briefed and shown films that linked priests and nuns to so-called subversives Bolasco and Yu 1981, p. 130). Many Filipinos still continued to join the Basic Christian Community alliance against social injustices, risking their lives in the process, even after the popular people power revolution overthrew the Marcos regime, which lasted from 1965 to 1986. For example, youths joined a cultural dance and theatrical troop called the Progressive Religious Association of Youth, Enlightenment, and Redemption (PRAYER, which served as the cultural arm of the Basic Christian Community movement), which provides free educational entertainment for strikers, demonstrators, victims of human rights abuses, and Basic Christian Community members in distant barrios. Many were illegally arrested, tortured, and framed by the subsequent Aquino administration for alleged crimes such as murder and robbery (Nadeau 2002). For this reason, the next President Ramos's declaration of political amnesty did not affect the status of these detainees, charged for common crimes (Nadeau 1995, p. 64). This practice of harassing, arresting, framing, and summarily executing often innocent civilians, including those who visibly question the social order to stand up for their rights—the right to own the land that they farm and for which they often have a legal title—continues, as of 2016, under the guise of President Rodrigo Duterte's (May, 2016—present) *shock and awe* policy that orders his military and policemen to shoot to kill whoever they perceive to be criminals. Some 1800 people were murdered by his (paramilitary) 'death squads' and military and policemen, during his first 4 months in office, and killings and civilian casualties, caught in the crossfires, in Duterte's war on crime and drugs are on the increase (Villamor and Paddock 2016).

Philippine Basic Ecclesial Communities, typically, share an ideology based on environmentalism, human rights, justice, peace, and grassroots democracy. Participants may seek solutions to problems through collective prayer and Scripture study, collective planning and decision-making, and negotiating with employers, government officials, development workers, and landlords. Local Basic Christian Community Organizers and practitioners of liberation theology, during the heyday of the overthrow of the Marcos dictatorial regime (1965–1986), on the heals of the fall of the Soviet Union (1922–1991), and in the face of on-going everyday violence, tend to oppose those who would dogmatically apply Marx's ideas as politics, rather than as a springboard for coming up with some new and practical ideas to solve local social and environmental problems from a bottom-up perspective.

Liberation theology, like all theology, is talk about God/Goddess. It is an inductive methodology and process that discerns a divine presence in the life of the people by taking into consideration their aspirations and then looking to see what the Bible has to say about that. Liberation theology is biased for the poor and oppressed because the God of the Bible is on the side of the poor who comes down to live with them. Liberation theology does not begin through the entry point of any absolute perspective, rather it goes back to the people to think and reflect upon their experiences and realities, to better understand and identify their problems in solidarity with the people concerned. Liberation theology is 'God/dess Talk' and its tools are largely based on hermeneutics (interpretation and real world applications of the spirit of Jesus's teachings) and Marxist analysis.

According to Paul Ricoeur (1979), hermeneutics can be defined as the study of symbols, which find their origin in some pre-linguistic bios, rather than culture or convention. He refers to *bios* as energy or desire in Freudian terms, and to the sacred in religious language. Ulin (1984) further explained that Ricoeur seeks to transcend a positivist tradition that divides nature from culture by rethinking them dialectically (pp. 105–109). Gadamer (1975), likewise, criticized Western positivism, which has developed an absolute notion of the science of reason. Ricoeur, slightly, diverges from Gadamer's view, however, in that he argues that symbols are not merely cultural constructions arrived at through inter-subjective consensus, rather they unite individuals to cosmic space. His method of looking at metaphors from this double view can be distinguished from Gadamer's theory of the meeting of horizons through discourse. Like Gadamer, Ricoeur argues that it is in the science of semiotics that human activity becomes objectified. Maurice Bloch (1977), by contrast, arrives at a somewhat different conclusion than that of Gadmaer and Ricoeur. He considers that it is within the relation between nature and culture that new conceptions are developed since they cannot come from a social structure defined as a shared system of meaningful categories. It is not in language-like processes that human activity becomes externalized. Bloch (1991) brought forth an alternative theory of connectivism to show how new ideas result from processes of associating visual and mental images with a rapidity faster than a mere sentence-logic model would allow. Despite individual differences and shortcomings, herme-neutics offers a guide for human decision-making and action because it provides a wide range of possibilities. In liberation theology, the term, hermeneutics, refers to human discernment of the spirit of biblical teachings. That is, Basic Christian Community participants interpret biblical readings by relating the scriptures to their own real world experiences and, then, asking what would Jesus do?

Welch (1985, p. 24, 34) looked at the early literature on liberation theology in relation to Foucault's archaeology of knowledge theory and her own feminist lens. She argued in a completely satisfying and substantial manner that liberation theolo-gies are not merely variant strains of thought within a traditional theology, for example, progressive versus conservative theologies within an overriding Catholic theology, rather they can be said to represent a completely new paradigm of knowl-edge, an epistemic break from traditional theology. They are continuous with only one tradition within Christianity as a whole, namely, with a tradition that is critical of social and religious hypocrisy.

As in Latin America, the ecological liberation theology movement in the Philippines earlier emerged in reaction to a form of development ideology associ-ated with authoritarian dictatorships (in the Philippine context the dictator Ferdinand Marcos). It represented a new revolutionary movement transcending class-based party politics and hierarchical organization in order to build an alternative system of power represented by the people (Nadeau 2002; Holden 2009). Liberation Theology is engaged in reconstructing modernization processes from the vantage point of victims: the poor and those who have been displaced by state led development and, more recently, climate change related natural disasters. It is a radical theology iden-tifying with the early Christian Church extant around the time of Christ, as Christ criticized the hypocrisy of the Scribes and rulers of his day. Pieres (1988) explains that Liberation theologians and their followers are oriented around the poverty and ethics of Christ; they also identify with other world religions (such as Hinduism, Buddhism, and Islam) claiming to have a message liberating the poor in so far as these religions also claim to be saviours of the commonweal. Adept religious lead-ers in such faiths voluntarily take a vow of poverty serving a double purpose as a powerful spiritual and political weapon against religious and secular leaders who act in a greedy and selfish manner. Armstrong (2008), founder of the Global Charter of Compassionate Communities argues that any current ideology (political, eco-nomic, religious, cultural or environmental) that does not present a sense of global understanding and compassion is failing. Ecological and feminist liberation theolo-gians, like Buddha, call everyone to feel with the other and encourage people to respect and live in harmony with each other and nature, in the presence of all that is sacred in every living thing. It is, after all, states Armstrong, by dethroning our-selves from the center and putting others there that we touch the divine. Secondly, local community based eco-liberation theology practitioners and their community partners identify with predecessors such as Christian friars and ministers who spoke out against social injustices under Spanish colonialism and, later, American colo-nialism. These church workers believe that they can change the structural roots of local poverty and create a more equitable, just, and environmentally green society, by working to rebuild communities affected by calamities, into more resilient and resourceful communities of compassion.

References

Armstrong, K. (2008). *Building compassionate communities*. Section 16. Retrieved from ctb. ku.edu

Bloch, M. (1977). The past in the present and the present in the past. *Man, 12*, 278–292.

Bloch, M. (1991). Language and anthropology and cognitive science. *Man, 26*(2), 183–198.

Bolasco, M., & Yu, R. (1981). *Church-state relations*. Manila: St. Scholastica's Press.

Cacayan, B., & Miclat, A. (1991). *Let your heart be bold: A reflection paper on church workers and national security in the Philippines*. Hong Kong: Asian Center for Progress.

De La Torre, E. (1986). *Touching ground, touching root*. Manila: Socio-Pastoral Institute.

Fabros, W. (1988). *The Church and its social involvement in the Philippines, 1930–1972*. Quezon City: Ateneo de Manila Press.

Gadamer, H. G. (1975). *Truth and method*. New York: Seabury.

Gaspar, K. (1990). *A people's option to struggle for creation*. Quezon City: Clareton.

Gaspar, K., CSsR. (1991). Redemptorist Brother, personal interview by Kathleen Nadeau at Redemptorist Justice and Peace Desk in Cebu City.

Holden, W. N. (2009). Post modern public administration in the land of promise: The basic ecclesial community movement of Mindanao. *Worldviews: Environment, Culture, Religion, 13*(2), 180–218.

Ileto, R. (1979). *Payson and revolution, popular movements in the Philippines, 1840–1910*. Quezon City: Ateneo de Manila University Press.

Marx, K. (1972). *Capital* (Vol. I, II, and III). New York: International.

Nadeau, K. (1995). *Ecclesial community movement in Cebu: The Philippines*. A dissertation presented to Arizona State University in partial fulfillment of the requirements for the Degree, Doctor of Philosophy.

Nadeau, K. (2002). *Liberation theology in the Philippines: Faith in a revolution*. Westport: Praeger.

Pieres, A. S. J. (1988). *An Asian theology of liberation*. New York: Orbis Books.

Ricoeur, P. (1979). The model of the text: Meaningful action considered as a text. In P. Rabinow & W. Sullivan (Eds.), *Interpretive social science: A reader* (pp. 91–117). Berkeley: University of California Press.

Ulin, R. (1984). *Understanding cultures: Perspectives in anthropology and social theory*. Austin: University of Texas Press.

Villamor, F., & Paddock, R. (2016, August 22). Nearly 1,800 killed in Duterte's War, Philippine Police Tells Senators. *New York Times*.

Welch, S. (1985). *Communities of resistance and solidarity, a feminist theology of liberation*. New York: Orbis Books.

Youngblood, R. (1990). *Marcos against the church: Economic development and repression in the Philippines*. Ithaca: Cornell University Press.

Chapter 8
Ecological Liberation Theology and the Philippines

Throughout much of history, the Roman Catholic Church was aligned with the rich and powerful of society and, with a few notable exceptions, showed little, if any, concern about the poor and marginalized. The church traditionally justified this disregard for temporal matters by using an approach known as the "distinction of planes," which argued that there were two planes of existence: the sacred plane (the concern of the church) and the secular plane (the concern of secular society) (Smith 1975). Any potentially destabilizing influences emerging from a discussion of Jesus' love for the poor in the scriptures were blunted by making it abundantly clear that any poverty being referred to was *spiritual poverty* and not *material poverty* (Nangle 2004).

In the years following World War II, however, it became increasingly apparent to many in the church that the demands of the modern world required it to make changes or risk becoming irrelevant. Then, during the second Vatican Council (1962–1965), the church began to make these changes. Specifically, Vatican II began to lead to a shift in Catholic teaching away from a purely spiritual understanding of salvation and towards a greater commitment to work for social justice and to challenge, and change, unjust structures in society (Smith 1975). The repercussions of Vatican II "would reverberate throughout the Christian world and influence the lives of an entire generation" (Gaspar 2005, p. 82). In Latin America, a part of the world, which was to profoundly influence Catholic thought and action, the church responded rapidly to Vatican II (Kinne 1990). When Vatican II described the role of the church as service to *the world*, Latin American theologians placed this within the context of *their world* of "underdevelopment, poverty and oppression" and they quickly came to see this as a "green light for social involvement" (Foroohar 1986, p. 39). It was in this context that the *Confederación Episcopal Latina América* (Latin American Episcopal Confederation or CELAM) met in Medellin, Colombia in August of 1968 and denounced the injustices inherent in Latin American social and economic structures, and resolved to firmly place the moral weight of the church on the side of those seeking reforms to benefit the poor (Smith 1975).

© The Author(s) 2017
W. Holden et al., *Ecological Liberation Theology*, SpringerBriefs in Geography,
DOI 10.1007/978-3-319-50782-8_8

As the impacts of Vatican II began to be felt during the late 1960s, and early 1970s, members of the Latin American church engaged in activism with the poor, began to use the conscientization method developed by the Brazilian educator Paulo Freire in *comunidades eclesiales de base* (basic ecclesial communities) where the poor were taught to *see* the reality of their poverty, were taught how to *judge* their reality of exploitation, and were taught how to *act* to change their reality by acting to change the unjust social structures confronting them (Brackley and Schubeck 2002). These *comunidades eclesiales de base*, were defined by Berryman (1987, p. 64), as "small lay-led communities, motivated by Christian faith, that see themselves as part of the church and that are committed to working together to improve their communities and to establish a more just society."

Given the common colonial heritage shared by the Philippines with Latin America, it did not take long for developments in Latin America to become influential (Nadeau 2002; Holden 2009). The church on Mindanao, Philippines, in the late 1960s and early 1970s, was very much open to change with newly established dioceses filled with people who had moved from elsewhere in the Philippines (Kinne 1990; Picardal 1995). According to Brother Karl Gaspar, a Redemptorist brother based in Davao City (and a member of the Ecumenical Association of Third World Theologians), these dioceses were also staffed by young bishops who were receptive to new ideas and who needed priests; in many cases they turned to foreign missionaries, such as the Maryknoll missionaries, to staff their parishes (Gaspar 2006, pers. interview). Many of these missionaries had experience organizing, and working with, *comunidades eclesiales de base* in Latin America and, thus, acted as a diffusive force carrying concepts from Latin America to Asia (Kinne 1990). In the early 1970s, Maryknoll missionaries established a Basic Christian Community (later, called Basic Ecclesial Community or BEC) program in what was then the Prelature of Tagum and this program then spread throughout Mindanao (Kinne 1990). Over time, particularly after the declaration of martial law in 1972, the worsening depravations of the Marcos dictatorship (1972–1986) served to intensify the advocacy of the Church on the behalf of the poor. Eventually, with the 1991 Second Plenary Council of the Philippines (PCP II) the church, as a whole, made a commitment to becoming a "church of the poor" and the vehicle to be used for achieving this goal was the BEC (Second Plenary Council of the Philippines 1991).

The term *basic* refers to both the *size* and the *social location* of the BECs (Holden 2009). They are *small* communities, consisting of from 40 to 200 families organized on a parish-by-parish basis, and most BECs consist of the *small people*, the poor, and the marginalized (Mendoza 2005). Members of all social classes may join BECs but, overwhelmingly, they are a movement of the poor as middle class and upper class people tend to refrain from participating in them (Holden 2009). Father Amado Picardal is a Redemptorist priest and Dean of the St. Alphonsus Theologate in Davao City. Father Amado provides seminars on BEC organizing for the Catholic Bishops Conference of the Philippines and has found that it is "easier to organize BECs among poor people than among rich people as there is more of a community among poor people than among rich people; as people's socioeconomic status goes up, the walls become higher, both literally and figuratively" (Picardal 2007, interview).

The term *ecclesial* emphasizes the place of the BECs within the church (Picardal 1999). Until PCPII, the BECs were known as "Basic Christian Communities," or BCCs, but with PCPII the decision was made to rename them as "Basic Ecclesial Communities" (Picardal 1999, interview). They are a way of being a church that is realized, located, and experienced at the grassroots (Picardal 1999). A BEC would not be *ecclesial* if it were not "united organically with other BECs, with the parish, with the diocese, and with the universal church" (Mendoza 2005, p. 63).

The term *community* emphasizes the communitarian nature of the BECs (Picardal 1999). These are not "societies" or "associations" but are communities whose members live in close spatial, and social, proximity to each other and who regularly interact with each other (Picardal 1999). The role of the BECs as communitarian organizations stands in sharp contrast with the individualistic, selfish, privatized, and competitive style that marks modern western culture (Azevedo 1993, p. 638). The BECs, states Picardal (1999), are intended to be "the realization of the ideal of the Christian community as described in the Acts of the Apostles" (Acts 2: 42–47 and Acts 4: 32–35).

Three main activities are engaged in by BECs: liturgical, developmental, and transformative (or "liberational"). Liturgical activities tend to the spiritual needs of their members, developmental activities cater to the material needs of their members, and transformative activities are designed to transform Philippine society into a better place (Mendoza 2005; Nadeau 2002; Picardal 1995). This is a descriptive typology, and BECs are not formally designated as "liturgical," "developmental," or "transformative" (Holden 2009), although funding agencies need to become more aware of the differences between these three models and wary of government and Church officials assessments of them (Nadeau 1995, p. 142). A given BEC, at any point in time, may be engaging in varying amounts of the three different activities and, in many cases, the demarcation between different activities may be blurred. For example, the parish priest may attend the BEC and provide a mass in its chapel (an ostensibly "liturgical" activity) and then discuss the dangers of climate change, during the homily (a "transformative activity").

These three types of BECs, liturgical, developmental, and transformative (liberational), may be perceived as stages of growth (Picardal 1995). The most rudimentary activities consist of liturgical activities catering to the *spiritual needs* of the members. Then, as a BEC becomes more developed, it continues with meeting its members' spiritual needs and begins to accommodate the *material needs* of its members by performing a developmental function. Finally, the most sophisticated BECs meet both the spiritual and material needs of its members while also acting to *transform* Philippine society and thus *liberate* its members from unjust social structures. In some areas, however, some BECs federated themselves into a people's organization because their local parish leaders wanted them to attend to their spiritual and material needs by concentrating on livelihood projects like candle-making, food processing and handicrafts production. But their BEC members had reached a more sophisticated level of conscientization and organizing that they wanted to address more fundamental issues confronting their precarious life like security of land/housing tenure and disaster risk confronting the community. Their informal settlement located at the deep river bend of the Marikina River beside a private residential subdivision have been increasingly at risk to flooding disasters over the years. Threats of flooding and eviction have forced them to band together and organize themselves for disaster protection. They realized that they have to protect the environment to reduce the risk of flooding their communities. Their strong faith in an All-providing God through nature and the environment permeates their daily life as reflected in Ka Noli's his daily ritual: "I start my day praying the rosary, then a passage from the gospel; then a prayer to St. Pedro Poveda and his morning prayer; then a prayer to St. Escriba, then a prayer to St. Michael the Archangel. After a simple breakfast, nurture the plants with homemade liquid fertilizer, to be assured of abundant flora to serve , in our little ways, as carbon sink of urbanized communities proximate to us, like Quezon City, Marikina City, and to a certain extent, Antipolo City" (Source: Interview of Ka Noli, longtime leader of Buklod Tao, Inc. in Banaba, San Mateo, Rizal).

In the early 1970s, as the BECs were being established on Mindanao, it had a migrant population that was more open to change; Mindanao's Christians had moved from elsewhere in the Philippines and, as they were willing to *move*, they became more receptive to *new things* (Holden 2009). This is consistent with two observations made by Hoffer (1963), in his seminal book *The True Believer: Thoughts on the Nature of Mass Movements*. First, emigration offers change and a chance for a new beginning; this means that those who are likely to move are also the same people who are wont to join a social movement (Hoffer 1963). Second, a social movement is much more likely to attract followers when people have "been separated by force or voluntarily from their hereditary milieu" (Hoffer 1963, p. 43). Another factor that has contributed to strong BECs in Mindanao is the historic use of missionary priests from religious orders (such as the Passionists or the Redemptorists) on that island (Picardal 1995). Members of religious orders are more likely to become involved in social action than diocesan priests and they have more institutional resources at their disposal than diocesan clergy, who depend more upon the resources of the diocese (Holden 2009). Attention now turns to an examination of how BECs in the Philippines, today, attempt to engage disaster relief work.

References

Azevedo, M. d. C. (1993). Basic ecclesial communities. In I. Ellacuria & J. Sobrino (Eds.), *Mysterium liberationis: Fundamental concepts of liberation theology*. New York: Orbis.

Berryman, P. (1987). *Liberation theology: Essential facts about the revolutionary movement in Latin America and beyond*. Philadelphia: Temple University Press.

Brackley, D., & Schubeck, T. L. (2002). Moral theology in Latin America. *Theological Studies, 63*(1), 123–160.

Foroohar, M. (1986). Liberation theology: The response of American catholics to socioeconomic problems. *Latin American Perspectives, 50*(13), 37–57. Summer.

Gaspar, K. M. (2005). *Mystic wanderers in the land of perpetual departure*. Manila: Institute of Spirituality in Asia.

Gaspar, K.M. (2006, January 5). Brother. *Redemptorist Brother*. Personal interview conducted by William Holden, Davao City, Philippines.

Hoffer, E. (1963). The true believer: Thoughts on the nature of mass movements. In *Time reading program* (Special ed.). New York: Time Incorporated.

Holden, W. N. (2009). Post modern public administration in the land of promise: The basic ecclesial community movement of Mindanao. *Worldviews: Environment, Culture, Religion, 13*(2), 180–218.

Kinne, W. (1990). *A people's Church? The Mindanao-Sulu Church Debacle*. Frankfurt am Main: Peter Lang.

Mendoza, F. (2005). *Basic ecclesial communities: The context and foundations of formation*. Mandaue City: Mandaue Printshop Corporation.

Nadeau, K. (1995). Ecclesial Community Movements in Cebu, The Philippines. Disseration for the Degree Doctor of Philosophy, Arizona State University, Tempe.

Nadeau, K. (2002). *Liberation theology in the Philippines: Faith in a revolution*. Westport: Praeger.

Nangle, J. (2004). *Birth of a Church*. Maryknoll: Orbis Books.

Picardal, A. L. (1995). Basic ecclesial communities in the Philippines: An ecclesiological perspective. Doctoral dissertation, Faculty of Theology, Pontifical Gregorian University, Rome.

Picardal, A. L., CSsR. (2007 and 2009). Interviewed at St. Alphonsus Theologate in Davao City, Philippines.

Second Plenary Council of the Philippines. (1991). *Acts and decrees of the second plenary council of the Philippines*. Manila: Catholic Bishops' Conference of the Philippines.

Smith, B. H. (1975). Religion and social change: Classical theories and new formulations in the context of recent developments in Latin America. *Latin American Research Review, 10*(2), 3–34.

Chapter 9
Philippine Basic Ecclesial Communities and Disaster Relief Work

Faith-based communities are very different from those motivated by competitive individualism. They often model themselves after the Basic Christian Communities, which are inspired by liberation theology and later by ecological theology and feminism. These rehabilitation and development communities have long engaged in a variety of livelihood projects designed to improve their members' quality of life. They are involved in income-generating activities such as handicraft production, food processing, garment making, soap making, cooperative stores, communal farming, and livestock dispersal programs (for more detailed examples, see Holden and Nadeau 2010 and 2011). They typically use a participatory approach that partners with local people who are interactively given an education on how to do research to determine what types of livelihood programs they would like to implement. In accordance with the Philippine's Local Government Code, they often help to monitor internal revenue allotments to local governments and the election process to make sure there is no corruption, but, as Evita Jimenez (2012) of the Center for People Empowerment explains, computerized voting technology makes this difficult because they have no way to make sure the requirements for transparency and integrity of the software have been complied with.

Women and men working with the basic communities are producing more varieties of vegetables and herbal medicines. Pharmaceutical drugs often are beyond the means of poor people unless they find medicines through groups that poor people do not always know about such as nonprofit, nongovernmental organizations or church clinics funded by outside and local donations. While public hospitals continue to offer free social and medical care for the poor, they remain understaffed and normally lack essential medical supplies. Patients have to bring their own sheets, towels, and beddings as well as any surgical instruments and medicines to be used. Sometimes, out of compassion and pity, doctors and nurses go into their own pockets to help the needy. Given the excessive cost of Western pharmaceuticals and health care, one important component of the basic community movement is to make traditional health care available to poor people by provisioning them with free

© The Author(s) 2017
W. Holden et al., *Ecological Liberation Theology*, SpringerBriefs in Geography,
DOI 10.1007/978-3-319-50782-8_9

herbal medicines for those who produce them, and selling them at inexpensive prices that local people can afford.

Philippine ecclesial communities engage in so many programs for women and the poor. For example, organic farming, solid waste management, and tree planting, the latter of which significantly helps reduce global warming. They contribute to community-based justice and education programs that increase their awareness about human rights, including women and LGBT (Lesbian, Gay, Bisexual, Transgender) equal rights.

In pre- and post-disaster communities, women's groups have been central in the prevention, mitigation, recovery and resilience building work. Women are viewed as the "guiding light" (*ilaw* in Tagalog or *suga* in Cebuano) in their families and households, more so during and after disasters. They prepare food for the children and pack their clothes for evacuation. A post-Ondoy/Ketsana study in Marikina City, Metro Manila showed that percentage-wise, female-headed households lost more (20% higher than male-headed ones) working days due to flooding, spent more time cleaning the premises and homes, washing clothes and attending to the children and sick members of the household (Porio 2014). After the Ondoy/Ketsana floods, several women volunteer health workers (Barangay Health Workers/BHWs) told the author (in Cebuano) during a community meeting: "Ma'am, our gender-sensitivity training told us that women carry a double burden (2 Bs), taking care of our families and looking for livelihood. *Ma'am, mula na nagbabaha dumami na ang B sa buhay namin dahil sa baha at hirap ng buhay* (Ma'am, since the flooding our Bs (burdens) have increased with the flooding and increasing economic difficulties. Nag-aalaga kami sa aming mga (we take care of our) *B*ata (children), *B*ana, *ang pinakamatanda kung bata* (Husband, my eldest child) *B*ahay (Home), **B**arangay (as BHWs), *B*asura (Solid Waste and recycling activities) and *B*aha (flood water)"! Indeed, climate-related hazards (typhoons, floods, sea level rise, and extreme heat/drought/rainfall) have made life for both men and women increasingly difficult, especially in poor, marginalized communities. But their social capital and trust networks provide them support and alleviate a bit their suffering (Porio 2016a, b).

Another interesting program of the Philippine ecological liberation movement is its work in the peace zones. According to Rufa Guiam of the Institute for Peace and Development at Mindanao State University in General Santos City, "decades of conflict have wrought substantial changes in women's lives, as well as male and female relationships." Only when women also participate equally and fully in negotiations for peace and justice can a truly sustainable peace be attained.

Women have long been at the forefront of negotiations between the Philippine government and the Moro National Liberation Front and other Muslim groups. On the signing of the Framework Agreement for Bangsamoro, in Mindanao, Philippines (October 2012), 150 soldiers and Filipino Christian men and women from the Basic Ecclesial Communities in the Peace Zone came together for the "Hijab Run for Peace: Religious Understanding Now" that was organized by the young Muslim professional network (Nadeau and Rayamajhi 2013, p. 40).

In Samar, a mass protest movement walked from Catarman, Borongan, and Basey to Catbalogan in resistance against the onslaught of mining operations, ransacking and wrecking havoc, in its last remaining forested zone. This well orga-nized and actively non-violent community based movement, largely, is credited for having given President Gloria Macapagal-Arroyo the impetus to issue Presidential Proclamation Number 442 on August 13, 2003, which created the Samar Island Natural Park. This park contains one of the largest contiguous remaining tracts of lowland tropical rainforest in the Philippines and it consists of a 3333 km^2 core area surrounded by a 1245 km^2 buffer zone providing an added layer of protection to the park as well as regulated benefits and livelihood opportunities for local communi-ties (Samar Island Biodiversity Project 2006).

The Samareños do not want new mining operations to interfere with the Natural Park and reduce the dampening effect it has on typhoons. In 2000, a survey was conducted of 1607 Samarnons and, when asked about their opinions of mining, 73% said they were opposed to it and the largest reason for this opposition, from 53% of all respondents, was that it destroys forest resources (Rosales and Francisco 2000). Large-scale mining will not help the poor and can make poor people even poorer by disrupting the environment upon which they depend (Holden and Jacobson 2012).

Relief workers in the ecumenical movement of the Philippines organize local faith-based communities that are team oriented to rebuild their lives and livelihoods. Millions of coastal populations have lost everything as a result of the impact of typhoons. Most notoriously, the most powerful storm on record named Yolanda, also called Super Typhoon Haiyan, hit the Philippines on November 8, 2013. Damage from Yolanda was estimated at more than US$830 million. Twenty-one fishing provinces and over 145,000 fishermen were affected. The storm negatively impacted populations in 471 municipalities and 51 cities.

The Philippine eco-theology and disaster relief movement at the intersections of the various and different cultural, religious, and atheistic human and environmental rights movements, reflects not only a strategy for more environmentally attuned reconstruction efforts that have the potential for better buffering and preventing future climate change related disasters from occurring, but the right of local com-munities to take charge of their decision making and individual-and cultural identities-and collectively inspired world views that differ from that of the ethos of capitalist individualism and globalization, coming down from above. But many organic intellectuals and local ecclesial community organizers believe that they can change the structural roots of poverty and create communities oriented toward jus-tice by working to change poor villages and neighborhoods into compassionate and collaborative self-help communities. Organic intellectual, a term coined by Gramsci (1988), as used here refers to working class intellectuals who listen to and, then, dialectically engage community participants in social action work to develop solu-tions to their most pressing problems by consensus building. Sometimes, a com-munity organizer's role is to work with local people, by holding workshops and networking to help individual members and the community, as a whole, better iden-tify and make the best possible use of all available resources that they may not have known about or how to access, otherwise. They effectively learn how the system works and how to better work the system, by being the change they seek to make.

Nathaniel Lerio, SSJV, of the Archdiocese of Cagayan de Oro, Northern Mindanao, explains, one of the important roles of the churches today is to facilitate community building for community resilience to climate change related disasters (Pers. Interview 2015; also, in Nadeau 2014). As he witnessed on December 17, 2011, carrying strong winds and rains, Typhoon Sedong hit Cagayan de Oro, directly, causing flash floods that destroyed many houses and settlements along its riverbanks. The Cagayan de Oro river serves as the basin of many river tributaries coming down from the nearby mountains and upland provinces of Bukidnon. Sendong wrecked havoc causing a tremendous amount of damage to houses and properties of informal city settlement communities. According to a report by the Office of Civil Defense, 35,000 families were badly affected, most of whom resided along the riverbanks. Many precious and valuable human lives were lost, and lined up along the way, as thousands of displaced families were forced to evacuate the areas.

In response, the Cagayan de Oro River Basin Council, sponsored by the local government, various agencies, and church funded NGOs, created a new initiative to reforest the neutered mountains and upper portions of the river basin that served as a watershed for the wider region. The council's role was to coordinate and bring together local government and national social welfare and development agencies, the Department of Energy and National Resources (DENR), and local and international NGOs such as Catholic Relief Services, (UN) Office of Coordinated Humanitarian Affairs, and Peace and Equity Foundation. These organizations participated in discussions to coordinate the emergency relief and later rehabilitation efforts.

The government identified lands for free housing with aid coming from local government and private institutions as San Miguel Corporation and the Filipino Chinese community that allowed the affected families to be resettled in Bukidnon centers, and other temporary housing shelters like those set up by the Catholic Relief Services. One thousand temporary shelters were built, with local bamboos and coconut lumbers, for displaced families waiting for more permanent shelters. After a period of time, the local government was able to provide 6000 more of these units but given the number of families displaced by Typhoon Sendong, which was more than 25,000, there remained a huge need for more housing and relocation sites.

Local diocesan church groups in coordination with the Catholic Bishops Council of the Philippines and the National Secretariat for Social Action received some support to augment local responses and were able to initiate more relocation sites, resettling more 190 families. Other Christian and non-Christian religious groups organized their own relief efforts and were able to accommodate 400 additional families. Families were given land and housing units on the basis of usufruct, meaning that they cannot sell or rent their property because, technically, it belongs to the government and other institutions.

There are still many families in need of relocation and more houses to be built. The local archdiocesan center is collaborating with the national government's Community Mortgage Program by partnering with local families to form socialized housing associations. Only organized Home Owners Associations (HOA) are eligible to apply for a government loan for social housing, which can be paid back in monthly installments, ranging from 100 to 800 pesos a month, over a 25 year period. Organized Home Owners Associations, in order to pre-qualify for government housing loans, are required to identify and negotiate the purchase price for the land and home constructions. The government's Community Mortgage Program offers a less expensive way for organized groups to purchase a home than the normal housing programs offered by private developers.

Community housing programs are community driven and, often, require facilitators to organize families in need, to help secure required materials for housing construction. They have to organize their own HOA Board of Directors and work together to obtain the necessary documents needed to process a government social housing loan. "Our role as a Church is to help facilitate this aspect of the process," states Father Lerio, by inviting families to gather together at community church centers and explaining how this cooperative housing program works. There is not enough relief funding left for the remaining thousands of families still waiting to be relocated to safe housing sites, by means of the usufruct system. So, we are organizing and empowering these families to work as teammates to rebuild their homes and communities, through their own efforts. To pre-qualify for the Community Mortgage Program, they must identify the land and negotiate a purchase price, with the land owner, who signs an agreement to wait until paid by the government, at which time the land will be turned over and taken possession of by the Home Owners Association. Instead of relying on the national government and others to dole out material supplies and provisions, these families are being empowered by social action workers to be able to organize on their own to approach the government and other providers of services to the community.

For example, in preparation for the 2013 local elections, some 20 Home Owners Associations, consisting of approximately 200 families, formed a united federation to discern how to actively participate in the upcoming democratic election process. They figured that unless they consciously worked for honest elections, they could not expect to have good leaders who would be a strong support for their relocation. They realized their active role and participated in the election process, by inviting the candidates to meet with them as a community. Visiting candidates were asked to sign a proposal, or social contract, written up by the participants, spelling out exactly what was expected of them should they win office. The participants, then, chose one candidate and actively campaigned on his/her behalf. Their candidate of choice won, and now they have a strong advocate in the local government.

In other words, explains Lerio, the role of the archdiocesan church of Cagayan de Oro was to motivate and help the poor and displaced families to realize their strength in being together. The church served by helping to facilitate the building and management of their organizations, as well as, doing mediation work when conflicts occurred. Its organized ecclesial communities were not exclusively Roman Catholic, but included Muslim and other non-Catholic members. "It is nice to see Muslims and Christians having their planning and evaluation meetings in the church centers," states Lerio. Community building is important for having resilient communities, and participation in good governance and environmentally sustainable development (Nadeau 2014, pp. 363–371).

That is, the struggle for social, economic, and environmental wellness is a long and gradual process of change. It is important for local government units and national and international governments and world organizations to empower outside community organizers and organic intellectuals to partner with local communities that are struggling to rebuild their homes and livelihoods, in the face of having been traumatically uprooted and displaced by climate change-related disasters. As Pope Francis, during his papal visit to the Philippines in January of 2015, suggested, we are all being called to immerse ourselves in the real world to begin the difficult work of rehabilitation and restructuring of society to solve the interrelated problem of poverty-environmental degradation-and climate-change.

References

Gramsci, A. (1988). Prison writings 1929–1935. In D. Forgacs (Ed.), *An antonio gramsci reader, selected writings* (pp. 1916–1935). New York: Schocken Books.

Guiam, R. (2012). Quoted in M. Buenaobra, With Framework Signed Women Walk the Road to Peace in Southern Mindanao in Asia, Weekly Insight and Analysis of the Asia Foundation. October 31, page 2. Also, see, Guiam R. and L. Dwyer (2012). Gender Conflict in Mindanao, Manila, Asia Foundation.

Holden, W., & Nadeau, K. (2010). Philippine liberation theology and social development in anthropological perspective. *Philippine Quarterly of Culture and Society, 38*(2), 89–129.

Holden, W., & Nadeau, K. (2011). Exemplifying accumulation by dispossession: Mining and indigenous peoples in the Philippines. *Geografiska Annaler (Swedish Society for Anthropology and Geography), 93*(2), 144–161.

Holden, W. N., & Jacobson, R. D. (2012). *Mining and natural hazard vulnerability in the Philippines: Digging to development or digging to disaster?* London: Anthem Press.

Jimenez, E. L. (2012). The hegemony of the culture of Philippine politics in Philippine elections. Paper presented at the International Conference on Philippine Studies, Kellog Hotel and Conference Center, East Lansing, Michigan, 28–30 Oct 2012.

Lerio, N., SSJV. (2015). Interviewd by Kathleen Nadeau at East Asian Pastoral Institute, Ateneo de Manila University, Loyola Heights, Metro Manila, Philippines, 20 Jan 2015.

Nadeau, K. (2014). Eco-theology and gender spirituality: Case for climate resilience in the Philippines. *East Asian Pastoral Review, 51*(4), 363–371.

Nadeau, K., & Rayamajhi, S. (2013). *Women's roles in Asia*. Santa Barbara: Greenwood.

Porio, E. (2014). Climate change adaptation in Metro Manila: Community risk assessment and power in community interventions in community interventions. In J. Fritz & J. Rheume (Eds.), *Community intervention, clinical sociology perspectives* (pp. 149–166). Netherlands: Springer Publications.

Porio, E. (2016a, July 10–15). *Risk and resilience in a rapidly unfolding world*. Closing Plenary Address, International Sociological Association, Vienna Forum.

Porio, E. (2016b, February 20). *Social capital, trust networks, recovery and resilience building in Post-Haiyan Communities (Samar, Leyte, Cebu)*. Paper presented before the "Post-Haiyan services and reflections on their effectiveness", Manila Observatory and Zuellig Foundation.

Rosales, R. M., & Francisco, H. A. (2000). *Estimating non-use values of the Samar Island Forest Reserve*. Washington: United States Agency for International Development.

Samar Island Biodiversity Project. (2006). *Final report of the terminal evaluation mission (April)*. United Nations Development Program.

Chapter 10
Conclusion

Putnam (1995), in his classic study, *Bowling Alone in America*, argued that the major institution that builds social capital and trust among community members is affiliation with their local churches or congregation. This observation resonates very well with the vibrant Philippine civic engagements of members in a Basic Ecclesial Community. Citizen's social capital and trust networks have been proven to be crucial in post-disaster recovery. In Chile, Fuster et al. (2015) demonstrated that local communities with strong social capital and trust networks, who have insider knowledge of the tectonic plates and movement of the place, can organize themselves better to protect themselves and attend to their immediate needs but this was later disregarded by external government agencies, during post reconstruction work. Women organized themselves, and were trusted to open food and medical reserves in the school, by the mayor, who trusted them to distribute it to families most in need. Yet, after the earthquake, the Chilean post-disaster agencies and decision-makers ignored this internal capacity and power of women and men, and their social networks that they could have organized, so the government suffered challenges.

Top-down reconstruction may cause division within a community by marginalizing organized groups and their knowledge and expertise. As already discussed, outside agencies and funding poured into the Philippines, after it was hit by a series of devastating typhoons. Yet, aid used to bring in temporary housing structures sometimes only caused more suffering as not everyone was given shelter and many had to wait long periods for more permanent housing. The bottom up Basic Ecclesial Community, in Cagayan de Oro, illustrated, by contrast, that local organizers can help to facilitate and empower traumatized people in the affected communities to work together to build a better life for themselves. They were able to produce well-constructed, good quality, and permanent homes within a reasonable time period.

In July of 2014, after typhoon Glenda hit Albay, Bicol region, unleashing horrific floods and landslides that covered and dislocated entire communities, taking a bottom up approach, the *European Union Humanitarian Aid* workers came to invest time and energy into helping the local people to prepare for extreme typhoons,

© The Author(s) 2017
W. Holden et al., *Ecological Liberation Theology*, SpringerBriefs in Geography,
DOI 10.1007/978-3-319-50782-8_10

floods, and other possible natural and man made hazards in the future (see, https://youtu.be/DDZNm4KQE9o). Their objective was to get everybody, local governing offices and barangay communities, near and far, involved. Men and women in the local barangays were individually approached, by local community organizers, and were given meaningful roles of responsibility. They formulated a Barangay Risk Reduction plan by brainstorming together and using local knowledge of the place. For example, the local government in Western Samar has no satellites for GPS mapping, so, instead, the local barangay community members did a *transect walk* to make a detailed map of the region. They came up with an indigenous warning system: communities in danger would warn other communities by blowing through a large conch shell, rather than ringing a bell. Most barangay members in these coastal communities relied on their boats and fishing to survive. They recognized and worked together to nurture overgrown trees, along coastal bays and shores, which provided their boats with protective natural covering, and enriched mangroves and seaside environments. As one of the local community members explained, "you don't need politicians to get things done, you can help without a position, by encouraging others." That is, where organizers invest their time and labor it can make a difference.

New millennial Philippine Basic Ecclesial Communities continue to struggle against top-down development and globalization processes by counter-posing bottom-up solutions that identify the interdependent relationship between culture and nature, especially in relation to relief and reconstruction work going on at the small agricultural and fishing village levels as well as in urban poor communities (Porio 2016). They are concerned with issues of gender, class, culture, literacy, and human- and environmental rights. Community organizers work with communities in impoverished urban and rural neighborhoods, war and conflict zones, and in areas hit hard by natural disasters to interactively and collaboratively brainstorm to find ways to provide needed housing in a safe and healthy environment, nutritional food supplies, medical services, and job security. They support collaborative relationships born of mutual respect for the dignity of the other, rather than on individual competition-and profit making (Reuther 2012, p. 1).

In the Philippines, as this brief demonstrated, because of government neglect and inability to respond to the needs of disaster-stricken communities, urban-rural poor communities have organized themselves into community-based organizations and/or peoples organizations working closely with their parishes and other faith-based organizations (FBOs). Frequent typhoons and flooding disasters, droughts and rice shortage have led them to organize or activate their community-based disaster risk reduction management councils (CB-DRRMC). Meanwhile women's groups have increasingly assumed community management roles in building their livelihood bases and provide support. Because of the inadequate government services for poor and marginalized groups, they have learned that they only have their own social capital trust networks to provide support before, during, and after disasters. But the series of climate disasters have transformed the structure of social capital/trust networks and social insurance among impoverished fishers, farmers and workers. Before Ketsana/Ondoy in 2009 and the series of flooding disasters caused the

Southwest Monsoon (Habagat) flooding in subsequent years, they have now increasingly relied on their local governments' disaster risk reduction management councils, evacuation centers, and relief operations. But such help often only lasts for a few days. In the end, they have to rely on their own social capital/trust networks with their neighbors and faith-based individuals/organizations to augment their daily needs/services and livelihood. Thus, the construction of community-based disaster recovery and resilience movement by the basic ecclesiastical communities (BECs) alongside humanitarian NGOs within the regulatory framework of government's relief and rehabilitation agencies are often fraught with contestations but laced with compromises, accommodations and alternative formulations along cultural, religious, political-economic and environmental lines.

References

Fuster, X., Imilan, W., & Vergara, P. (2015). Post-disaster reconstruction without citizens and their social capital in Llico, Chile. *Environment and Urbanization*, 27, 317–326.

Porio, E. (2016, July). Risk and resilience in a rapidly unfolding world. Closing Plenary Address, International Sociological Association, Vienna Forum.

Putnam, R. D. (2000). *Bowling alone: The collapse and revival of American community*. New York: Simon & Schuster paperbacks.

Reuther, R. (2012). *Ecofeminism*. Retrieved from http://www.spunk.org/texts/pubs/openeye/sp000943.txt

Index

© The Author(s) 2017
W. Holden et al., *Ecological Liberation Theology*, SpringerBriefs in Geography,
DOI 10.1007/978-3-319-50782-8